SCHAUM'S™
EASY OUTLINES

Elementary Algebra

Online Diagnostic Test

Go to **Schaums.com** to launch the Schaum's Diagnostic Test.

This convenient application provides a 30-question multiple-choice test that will pinpoint areas of strength and weakness to help you focus your study. Questions cover all aspects of elementary algebra, and the correct answers are explained in full. With a question-bank that rotates daily, the Schaum's Online Test also allows you to check your progress and readiness for final exams.

Other titles featured in Schaum's Online Diagnostic Test:

Schaum's Easy Outlines: Calculus, 2nd Edition
Schaum's Easy Outlines: Geometry, 2nd Edition
Schaum's Easy Outlines: Statistics, 2nd Edition
Schaum's Easy Outlines: College Algebra, 2nd Edition
Schaum's Easy Outlines: Biology, 2nd Edition
Schaum's Easy Outlines: Human Anatomy and Physiology, 2nd Edition
Schaum's Easy Outlines: Beginning Chemistry, 2nd Edition
Schaum's Easy Outlines: Organic Chemistry, 2nd Edition
Schaum's Easy Outlines: College Chemistry, 2nd Edition

Elementary Algebra

—————————— *Second Edition*

Barnett Rich, Ph.D.
Revised by Philip A. Schmidt, Ph.D.

Abridgement Editor:
Brenda Bradford

New York   Chicago   San Francisco   Lisbon   London   Madrid   Mexico City
Milan   New Delhi   San Juan   Seoul   Singapore   Sydney   Toronto

The *McGraw·Hill* Companies

1 2 3 4 5 6 7 8 9 10 11 12 13 14 15   WFR/WFR   1 9 8 7 6 5 4 3 2 1 0

ISBN   978-0-07-174583-3
MHID   0-07-174583-1

Library of Congress Cataloging-in-Publication Data

Moyer, Robert E.
    Schaum's easy outline of elementary algebra / Robert Moyer. — 2nd ed.
       p.   cm. — (Schaum's easy outline)
    Includes index.
    ISBN 0-07-174583-1 (alk. paper)
    1. Algebra—Outlines, syllabi, etc.   2. Algebra—Problems, exercises,
etc. I. Title. II. Title: Elementary algebra.

   QA159.2.C69   2010
   512.9—dc22   2010011634

McGraw-Hill books are available at special quantity discounts to use as premiums and sales promotions or for use in corporate training programs. To contact a representative, please e-mail us at bulksales@mcgraw-hill.com.

This book is printed on acid-free paper.

Contents

Chapter 1
FROM ARITHMETIC TO ALGEBRA

Representing Numbers by Letters

In this chapter, we are going to lead you from arithmetic to algebra. Underlying algebra as well as arithmetic are the four fundamental operations:

1. Addition (sum)
2. Subtraction (difference)
3. Multiplication (product)
4. Division (quotient)

In algebra, **letters may be used to represent numbers.** By using letters and mathematical symbols, short algebraic statements replace lengthy verbal statements.

Verbal Statement	Algebraic Statement
Seven times a number reduced by the same number equals six times the number	$7n - n = 6n$

When a number is multiplied by a letter, the multiplication sign may be omitted.

Problem 1.1
State the product below without using a multiplication sign:

a) $7 \times y$ *Ans.* a) 7y

Problem 1.2
Replace the verbal statement with an equivalent algebraic equation:

a) If 6 times a number is reduced by the same number, the result must be 5 times the number.

Ans. a) $6n - n = 5n$

Interchanging Numbers in Addition and Multiplication

Interchanging Numbers in Addition

Addends are numbers that are being added. In $5 + 3 = 8$, the addends are 5 and 3. Addends may also be letters representing numbers. These are known as *literal addends*. Thus in $5 + a = 8$, the numerical addend is 5 and the literal addend is *a*.

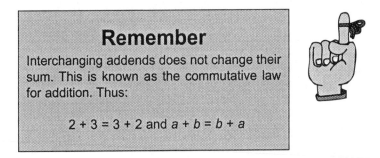

Remember
Interchanging addends does not change their sum. This is known as the commutative law for addition. Thus:

$$2 + 3 = 3 + 2 \text{ and } a + b = b + a$$

Interchanging addends may be used to help simplify addition.

Example 1.1
Simplify the addition by interchanging addends:

   a) $20 + 73 + 280$

Ans. a) $20 + 280 + 73$
        $300 + 73 = 373$

Interchanging Numbers in Multiplication

Factors are numbers being multiplied. Thus, in $5 \times 3 = 15$, the factors are 5 and 3. Factors may also be letters representing numbers. These are known as **literal factors**. Thus in $5 \times a = 15$, the numerical factor is 5 and the literal factor is a.

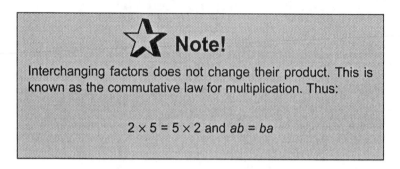

Interchanging factors does not change their product. This is known as the commutative law for multiplication. Thus:

$$2 \times 5 = 5 \times 2 \text{ and } ab = ba$$

Interchanging factors may be used to help simplify multiplication.

Example 1.2
Simplify the multiplication by interchanging factors:

    a) $25 \times 19 \times 4 \times 2$

Ans. a) $25 \times 4 \times 19 \times 2$
            $100 \times 38 = 3800$

Expressing Operations Algebraically

Symbolizing the Operations in Algebra

The symbols for the fundamental operations are as follows:

    1. Addition: +

    2. Subtraction: −

3. Multiplication: ×, ( ), ·, no sign

4. Division: ÷, :, fraction bar

Thus, $n + 4$ means "add n and 4." $4 \times n$, $4(n)$, $4 \cdot n$, and $4n$, mean "multiply n and 4." $n - 4$ means "subtract 4 from n." $n \div 4$, n: 4, and $\dfrac{n}{4}$ mean "n divided by 4."

Problem 1.3
Symbolize each, using multiplication signs:

    a) 8 times 11
    b) b times c

Ans. a) 8×11, $8 \cdot 11$, $8(11)$, or $(8)(11)$
    b) $b \cdot c$ or bc (avoid $b \times c$)

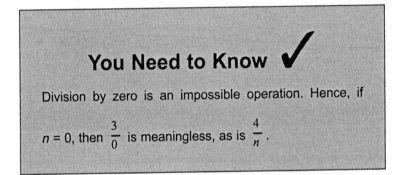

You Need to Know ✔

Division by zero is an impossible operation. Hence, if $n = 0$, then $\dfrac{3}{0}$ is meaningless, as is $\dfrac{4}{n}$.

Problem 1.4
When is each division impossible?

    a) $\dfrac{10}{b}$ b) $\dfrac{8}{x-5}$

Ans. a) if $b = 0$ b) if $x = 5$

Expressing Addition and Subtraction Algebraically

In algebra, changing verbal statements into algebraic expressions is of major importance. The operations of addition and subtraction are denoted by words such as the following:

Words Denoting Addition		Words Denoting Subtraction	
Sum	More than	Difference	Less than
Plus	Greater than	Minus	Smaller than
Gain	Larger than	Less	Fewer than
Increase	Enlarge	Decrease	Shorten

Problem 1.5

If n represents a number, express algebraically:

    a) the sum of the number and 7
    b) 25 less than the number

Ans.   a) $n + 7$ or $7 + n$
       b) $n - 25$

Problem 1.6

Express the statement below algebraically:

    a) a price 60 dollars cheaper than p dollars

Ans.   a) $p - 60$

Expressing Multiplication and Division Algebraically

Knowing how to change verbal statements involving multiplication and division into algebraic expressions is also very important. These operations are denoted by words such as the following:

Words Denoting Multiplication		Words Denoting Division	
Multiplied by	Double	Divided by	Ratio
Times	Triple	Quotient	Half
Product	Quadruple		
Twice	Quintuple		

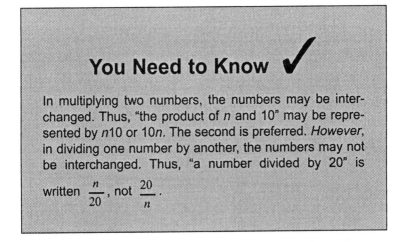

You Need to Know ✔

In multiplying two numbers, the numbers may be interchanged. Thus, "the product of n and 10" may be represented by n10 or 10n. The second is preferred. *However*, in dividing one number by another, the numbers may not be interchanged. Thus, "a number divided by 20" is written $\dfrac{n}{20}$, not $\dfrac{20}{n}$.

Problem 1.7
What statements may be represented by the operation below?

a) $\dfrac{5w}{7}$

Ans. a) 1. five-sevenths of w
2. 5w divided by 7
3. quotient of 5w and 7
4. ratio of 5w to 7

Expressing Two or More Operations Algebraically

Parentheses ( ) are used to treat an expression as a single number. Thus, to double the sum of 4 and x, write $2(4 + x)$.

Problem 1.8

Express algebraically:

a) a increased by twice b *Ans.*   a) $a + 2b$

b) 30 decreased by 3 times c b) $30 - 3c$

c) the average of s and 20   c) $\dfrac{s + 20}{2}$

d) two-thirds the sum of   d) $\dfrac{2}{3}\left(n + \dfrac{3p}{7}\right)$
$\quad$ n and three-sevenths of p

Order of Operations

In evaluating or finding the value of an expression containing numbers, the operations involved must be performed in a certain order. In expressions not containing parentheses, **multiplication and division always precede addition and subtraction.**

Example 1.3

Evaluate the following numerical expression:

a) $3 + 4 \times 2$

Ans.   a) 1. Do **multiplication** $3 + 4 \times 2$
$\quad\quad\quad$ **and division** in order
$\quad\quad\quad$ from left to right: $3 + 8$

2. Do remaining **addition**
   and subtraction in order
   from left to right: *Ans.* 11

The same rules and procedures hold true for algebraic expressions as well. The first step in evaluating algebraic expressions is to substitute for each given letter. Note the following example:

Example 1.4
Evaluate the following algebraic expression:

a) $x + 2y - \dfrac{z}{5}$ when $x = 5$, $y = 3$, $z = 20$.

Ans.   a) 1. **Substitute** the value
            given for each letter:

$$x + 2y - \frac{z}{5}$$

$$5 + 2(3) - \frac{20}{5}$$

2. Do **multiplication and**
   division from left to right: $5 + 6 - 4$
3. Do remaining **addition and**
   subtraction in order from left to right: *Ans.* 7

Essential Point

Because parentheses may be used to treat an expression as a single number, they may change the order of operations. Thus, to evaluate 2(4 + 3), add 4 and 3 in the parentheses before multiplying.

Example 1.5

Evaluate the following algebraic expression containing parentheses:

a) $2(a + b) + 3a - \dfrac{b}{2}$ if $a = 7$ and $b = 2$.

Ans. a) 1. Substitute the value given for each letter:

$$2(a + b) + 3a - \dfrac{b}{2}$$

$$2(7 + 2) + 3(7) - \dfrac{2}{2}$$

2. Evaluate inside parentheses: $2(9) + 3(7) - \dfrac{2}{2}$
3. Do multiplication and division in order from left to right:

$$18 + 21 - 1$$

4. Do remaining additions and subtraction in order from left to right:

Ans. 38

Repeated Multiplying of a Factor: Base, Exponent, and Power

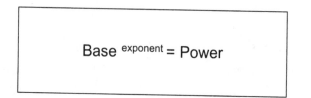

Base ^{exponent} = Power

In $2 \cdot 2 \cdot 2 \cdot 2 \cdot 2$, the factor 2 is being multiplied repeatedly. This may be written in a shorter form as 2^5 where the repeated factor 2 is the **base** while the small 5 written above and to the right of 2 is the **exponent**. The answer 32 is called the fifth **power** of 2.

An exponent is a number that indicates how many times another number, the base, is being used as a repeated factor. Thus, in the equation $b \cdot b$ or $b^2 = 81$, b is the base, 2 is the exponent, and 81 is the second power of b.

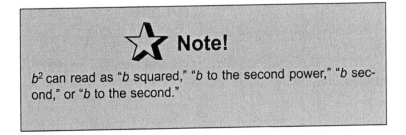

Note!

b^2 can read as "b squared," "b to the second power," "b second," or "b to the second."

Problem 1.9
Write each, using bases and exponents:

a) $3 \cdot 3 \cdot 7 \cdot 7$ b) $bbbbb$ c) $2(a + b)(a + b)$

Ans.   a) $3^2 7^2$ b) b^5 c) $2(a + b)^2$

Problem 1.10
Write each without using exponents:

a) $5 \cdot 7^3 \cdot 8$ b) $6(5y)^2$

Ans.   a) $5 \cdot 7 \cdot 7 \cdot 7 \cdot 8$ b) $6(5y)(5y)$

Problem 1.11
Evaluate the following:

a) $\dfrac{1}{2} \cdot 2^4 \cdot 3^2$ b) $(3 + 4^2)(3^3 - 5^2)$

Ans.   a) $\dfrac{1}{2} \cdot 16 \cdot 9 = 72$ b) $19 \cdot 2 = 38$

Problem 1.12
Evaluate if $a = 5$, $b = 1$, and $c = 10$:

a) $(2a)^2$ b) $(c + 3b)^2$

Ans. a) $10^2 = 100$ b) $13^2 = 169$

Combining Like and Unlike Terms

Like terms or similar terms are terms having the same literal factors, each with the same base and same exponent. Thus:

Like Terms	Unlike Terms
$7x$ and $5x$	$7x$ and $5y$
$8a^2$ and a^2	$8a^2$ and a^3

Remember that the exponent applies only to the number immediately in front of it.

Example 1.6
Combine the following:

a) $7x + 5x - 3x$

Ans. a) 1. Add or subtract numerical $7x + 5x - 3x$
coefficients: $7 + 5 - 3$
$= 9$
2. Keep common literal factor: *Ans.* $9x$

Problem 1.13
Simplify each expression by combining like terms:

a) $18a + 12a - 10$
b) $6b + 20b + 2c - c$

Ans.   a) $30a - 10$
          b) $26b + c$

Signed Numbers

Symbolizing the Operations in Algebra

Signed numbers are positive or negative numbers used to represent quantities that are opposites of each other. Thus, if +25 represents 25° above zero, −25 represents 25° below zero.

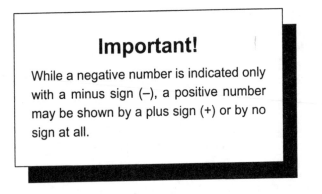

Important!

While a negative number is indicated only with a minus sign (−), a positive number may be shown by a plus sign (+) or by no sign at all.

The table below illustrates pairs of opposites that may be represented by +25 and −25.

+25	−25
$25 deposited	$25 withdrawn
25 lb gained	25 lb lost
25 mi to the north	25 mi to the south

The absolute or numerical value of a signed number is the number that remains when the sign is removed. Thus, 25 is the absolute value of +25 or −25.

Problem 1.14
State the quantity represented by each signed number:

  a) by −10, if +10 means 10 yards gained
  b) by −5, if +5 means $5 earned
  c) by +15, if −15 means 15 miles south

Ans. a) 10 yards lost
     b) $5 spent
     c) 15 miles north

Adding Signed Numbers

In algebra, adding signed numbers means combining them to obtain a single number that represents the total or combined effect. Remember these three rules for adding signed numbers.

• *Rule 1* — To add two signed numbers with like signs, add their absolute values. To this result, prefix the common sign. Thus, to add +7 and +3 or to add −7 and −3, add the absolute values 7 and 3. To the result 10, prefix the common sign. Hence, $+7 + (+3) = +10$ and $-7 + (-3) = -10$.

Example 1.7
Add the following numbers with like signs:

  a) +8, +2

Ans. a) 1. Add absolute values: $+8 + (+2)$
                               $8 + 2$
                               $= 10$

     2. Prefix common sign: *Ans.* +10

• *Rule 2* — To add two signed numbers with unlike signs, subtract the smaller absolute value from the other. To this result, prefix the sign of the number having the larger absolute value. Thus, to add +7 and −3 or to −7 and +3, subtract the absolute 3 from the absolute value 7. To the result 4, prefix the sign of the number having the larger absolute value. Hence, +7 + (–3) = +4 and –7 + (+3) = –4.

Example 1.8
Add the following numbers with unlike signs:

 a) +7, –5

Ans. a) 1. Subtract absolute values: $+7 + (-5)$
$$7 - 5$$
$$= 2$$

 2. Prefix sign of number having
   larger absolute value: Sign of 7 is +
                                        Ans. +2

• *Rule 3* — Zero is the sum of two signed numbers with unlike signs and the same absolute value. Such signed numbers are **opposites** of each other. Hence, +27 + (–27) = 0.

Example 1.9
Add the following signed numbers that are opposites:

 a) –18, +18

Ans. a) Sum is always zero: $-18 + (+18)$
$$= 0 \quad Ans.$$

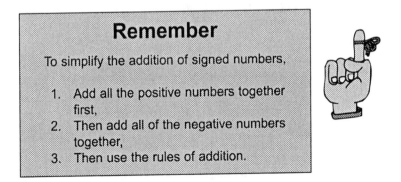

Remember

To simplify the addition of signed numbers,

1. Add all the positive numbers together first,
2. Then add all of the negative numbers together,
3. Then use the rules of addition.

Subtracting Signed Numbers

The symbol, – , used in subtracting signed numbers, means both "subtract" and "negative number." Thus, +8 – (–15) means subtract negative 15 from positive 8.

- *Rule 1* — To subtract a positive number, add its opposite negative. Thus, to subtract +10, add –10. Hence, (+18) – (+10) becomes (+18) + (–10) which = +8.

Example 1.10
Subtract the following positive number:

a) +8 from +29

Ans. a) To **subtract** a postive, **add**
its opposite negative:
$$+29 - (+8)$$
$$+29 + (-8)$$
$$= 21 \ Ans.$$

- *Rule 2* — To subtract a negative number, add its opposite positive. Thus, to subtract –10, add +10. Hence, (+30) – (–10) becomes (+30) + (+10) which = +40.

Example 1.11
Subtract the following negative number:

a) –7 from +20

Ans. a) To **subtract** a negative, **add**
its opposite positive:
$$+20 - (-7)$$
$$+20 + (+7)$$
$$= 27 \ Ans.$$

Multiplying Signed Numbers

Multiplying Two Signed Numbers

• *Rule 1* — To multiply two signed numbers with like signs, multiply their absolute values and make the product positive. Thus, $(+5)(+4) = +20$ and $(-5)(-4) = +20$

Example 1.12
Multiply the following numbers with like signs:

    a) $(+5)(+9)$

Ans.   a) 1. Multiply the absolute values: $(+5)(+9)$
$$5(9)$$
$$= 45$$

        2. Make product positive: $+45$ *Ans.*

• *Rule 2* — To multiply two signed numbers with unlike signs, multiply their absolute values and make the product negative. Thus, $(-7)(+2) = -14$ and $(+7)(-2) = -14$

Example 1.13
Multiply the following numbers with unlike signs:

    a) $(+8)(-9)$

Ans.   a) 1. Multiply absolute values: $(+8)(-9)$
$$8(9)$$
$$= 72$$

        2. Make product negative: -72 *Ans.*

• *Rule 3* — Zero times any number equals zero. Thus, $(0)(-8) = 0$ and $(44.7)(0) = 0$.

Example 1.14
Multiply the following numbers:

  a) (+10)(0)

Ans.   a) Product is always zero: (+10)(0)
                                                = 0 *Ans.*

Multiplying More Than Two Signed Numbers

• *Rule 4 —*   Make the product positive if all the signed numbers are
            positive or there are an even number of negatives. Thus,
            (+10)(+4)(−3)(−5) = +600.

Example 1.15
Multiply the following numbers:

  a) (+2)(+3)(+4)

Ans.   a) 1. All signs are positive.   (+)(+)(+)

        2. Make the product positive
           if all the signed numbers are
           positive or there are an even
           number of negatives: +24 *Ans.*

• *Rule 5 —* Make the product negative if there are an odd number of
            negatives. Thus, (+10)(−4)(−3)(−5) = −600.

Example 1.16
Multiply the following numbers:

  a) (+2)(+3)(−4)

Ans.   a) 1. Count the negatives :   (+)(+)(−)
                                              Odd number of negatives

    2. Make the product negative
       if there are an odd number
       of negatives : -24 *Ans.*

• *Rule 6* — Make the product zero if any number is zero. Thus, $(-5)(+82)(0)(316) = 0$.

Example 1.17
Multiply the following numbers:

      a) $(-1)(-2)(+5)(+10)(0)$

Ans.   a) Product is always zero: $(-1)(-2)(+5)(+10)(0)$
$$= 0\quad Ans.$$

Using Signed Numbers To Solve Problems

Example 1.18
Complete the statement, using signed numbers to obtain the answer:

    a)   If Tracy deposits $5 each week, then after 3 weeks her bank balance will be what?

Ans. a)   Let $+5 = \$5$ weekly deposit
           And $+3 = 3$ weeks later
           Then $(+5)(+3) = +15$
           Her bank balance will be $15 more. *Ans.*

Finding Values of Signed Numbers Having Exponents

• *Rule 1* — If the base and exponent are positive, the value of the number is always positive. Thus, $(+2)^3$ or $2^3 = +8$ since $(+2)^3 = (+2)(+2)(+2)$.

Problem 1.15

Find the value:

    a) 3^2 *Ans.*   a) 9

• *Rule 2* — For negative bases having even exponents, the value is always positive. Thus, $(-2)^4 = +16$, since $(-2)^4 = (-2)(-2)(-2)(-2)$.

Problem 1.16

Find the value:

    a) $(-3)^2$ *Ans.*   a) 9

• *Rule 3* — For negative bases having odd exponents, the value is always negative. Thus, $(-2)^5 = -32$, since $(-2)^5 = (-2)(-2)(-2)(-2)(-2)$.

Problem 1.17

Find the value:

    a) $(-0.2)^3$ *Ans.*   a) –0.008

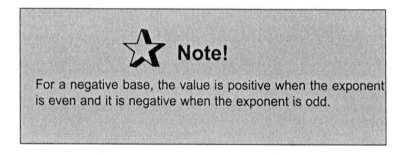

⭐ Note!

For a negative base, the value is positive when the exponent is even and it is negative when the exponent is odd.

Dividing Signed Numbers

• *Rule 1* — To divide two signed numbers with like signs, divide the absolute value of the first by that of the second and make

the quotient positive. Thus, $\dfrac{+8}{+2} = +4$ and $\dfrac{-8}{-2} = +4$.

Example 1.19
Divide:

a) +12 by +6   *Ans.*   a) $\dfrac{+12}{+6} = +2$

• *Rule 2* — To divide two signed numbers with unlike signs, divide the absolute value of the first by that of the second and

make the quotient negative. Thus, $\dfrac{+12}{-4} = -3$ and $\dfrac{-12}{+4} = -3$.

Example 1.20
Divide:

a) +20 by –5   *Ans.*   a) $\dfrac{+20}{-5} = -4$

• *Rule 3* — Zero divided by any number is zero. Thus, $\dfrac{0}{+17} = 0$ and $\dfrac{0}{-17} = 0$.

Example 1.21

Divide:

a) 0 by −12   *Ans.*   a) $\dfrac{0}{-12} = 0$

Combining Multiplying and Dividing of Signed Numbers

• *Rule 1* — Make the sign of the answer positive if all the numbers are positive or there are an even number of negatives.

Thus, $\dfrac{(+12)(+5)}{(+3)(+2)}, \dfrac{(+12)(+5)}{(-3)(-2)}, \dfrac{(-12)(-5)}{(-3)(-2)} = +10$.

Example 1.22

Divide:

a) $\dfrac{(+12)(+8)}{(-3)(-4)}$

Ans. a) $\dfrac{(+)(+)}{(-)(-)} = +$ answer, and $\dfrac{(12)(8)}{(3)(4)} = 8$, so $= +8$

• *Rule 2* — Make the sign of the answer negative if there is an odd

number of negatives. Thus, $\dfrac{(+12)(+5)}{(+3)(-2)}, \dfrac{(+12)(-5)}{(-3)(-2)},$

$\dfrac{(-12)(-5)}{(+3)(-2)} = -10$.

Example 1.23
Divide:

a) $\dfrac{(-4)}{(-5)(-10)}$

Ans. a) $\dfrac{(-)}{(-)(-)} = -$ answer, and $\dfrac{(4)}{(5)(10)} = \dfrac{2}{25}$, so $= -\dfrac{2}{25}$

• *Rule 3* — Make the answer zero if one of the numbers in the

numerator is zero. Thus, $\dfrac{(+53)(0)}{(-17)(-84)} = 0$

Problem 1.18
Solve:

a) $\dfrac{(+27)(0)}{(-3)(+15)}$ 　　　*Ans.*　a) 0

Evaluating Expressions Having Signed Numbers

To evaluate an expression having signed numbers:

1. Substitute the values given for the letters, always enclosing them in parentheses.

2. Perform the order of operatons, evaluating the numbers with exponents first.

Example 1.24
Evaluate if $y = -2$:

a) $4y^2$ *Ans.*   a) $4(-2)^2$

 $4(4)$

 16 *Ans.*

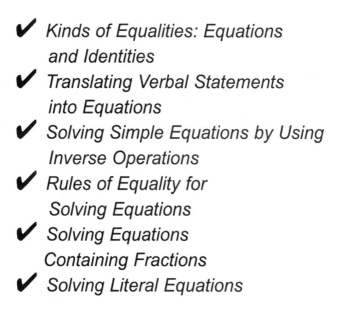

Chapter 2
SIMPLE EQUATIONS AND THEIR SOLUTIONS

IN THIS CHAPTER:

✔ *Kinds of Equalities: Equations and Identities*
✔ *Translating Verbal Statements into Equations*
✔ *Solving Simple Equations by Using Inverse Operations*
✔ *Rules of Equality for Solving Equations*
✔ *Solving Equations Containing Fractions*
✔ *Solving Literal Equations*

25

Kinds of Equalities:
Equations and Identities

An **equality** is a mathematical statement that two expressions are equal, or have the same value. Thus, $2n = 6$, $2n + 3n = 5n$, and $16 = 16$ are equalities. There are two types of equalities:

• **Equations** — an equation is an equality in which the unknown or unknowns may represent only a particular value or values. An equation is a conditional equality. Thus, $2n = 12$ is an equation since n may have only one value, 6.

• **Identities** — an identity is an equality in which the unknown or unknowns may represent any value. An identity is an unconditional equality. Thus, the equations $2n + 3n = 5n$ and $x + y = y + x$ are identities since there are no restrictions on the values that n, x, and y may have.

A **root of an equation** is any number which, when substituted for the unknown, will make both sides of the equation equal. A root is said to **satisfy the equation**. Thus, 6 is a root of $2n = 12$, while 5 or any other number is not. **Checking an equation** is the process of substituting a **particular value for an unknown to see if the value will make both sides equal.**

Example 2.1
By checking, determine which value is a root of each equation:

    a)   Check $2n + 3n = 25$ for $n = 5$ and $n = 6$

Ans.   a)   Check: $n = 5$ Check: $n = 6$

$$2n + 3n = 25 \qquad\qquad 2n + 3n = 25$$
$$2(5) + 3(5) =^? 25 \qquad 2(6) + 3(6) =^? 25$$
$$10 + 15 =^? 25 \qquad\qquad 12 + 18 =^? 25$$
$$25 = 25 \qquad\qquad\qquad 30 \neq 25$$

5 is a root of $2n + 3n = 25$. *Ans.*

Translating Verbal Statements into Equations

In algebra, a verbal problem is solved when the value of its unknown (or unknowns) is found. In the process it is necessary to "translate" verbal statements into equations.

Problem 2.1
Translate into an equation, letting n represent the number: (You do not need to solve the equation.)
    a) One–half of what number equals 10?
    b) Twice the sum of what number and 5 is 24?
    c) 4 less than what number is eight?

Ans.   a) $\frac{1}{2} n = 10$
        b) $2(n + 5) = 24$
        c) $n - 4 = 8$

Example 2.2
Represent the unknown by a letter and write an equation for each problem: (You do not need to solve the equation.)

    a) A woman worked for 5 hours and earned $20.75. What was her hourly wage?
    b) A baseball team won 4 times as many games as it lost. How many games did the team lose if it played a total of 100 games?

Ans.   a) Let w = hourly wage in dollars.
          Then, $5w = 20.75$. *Ans.*
       b) Let n = number of games lost and
          $4n$ = number of games won
          Then, $n + 4n = 100$. *Ans.*

Solving Simple Equations by Using Inverse Operations

Inverse operations are two operations such that if one is involved with the unknown in the equation, then the other is used to solve the equation. To solve the equation by using inverse operations, think of it as asking a question such as in the following equations:

Equation	Question Asked by Equation	Finding Root of Equation
$n + 4 = 12$	What number plus 4 equals 12?	$n = 12 - 4 = 8$
$\dfrac{n}{4} = 12$	What number divided by 4 equals 12?	$n = 12 \cdot 4 = 48$

The first equation above involves **addition** and is solved by **subtracting** 4 from 12. The second equation above involves **division** and is solved by **multiplying** 4 by 12.

★ Note!

Inverse operations are two operations such that if one is involved with the unknown in the equation, then the other is used to solve the equation:

1. Addition and subtraction are inverse operations
2. Multiplication and division are inverse operations

Example 2.3
Solve each equation:

a) $x + 3 = 8$ d) $3x = 12$

b) $5 + y = 13$ e) $12y = 3$

c) $x - 10 = 2$ f) $\dfrac{y}{12} = 3$

Ans. a) $x = 8 - 3$ or 5 d) $x = \dfrac{12}{3}$ or 4

b) $y = 13 - 5$ or 8 e) $y = \dfrac{3}{12}$ or $\dfrac{1}{4}$

c) $x = 2 + 10$ or 12 f) $y = 3 \cdot 12$ or 36

Rules of Equality for Solving Equations

There are four rules of equality for solving equations. They are as follows:

1. *Addition Rule of Equality* — To maintain an equality, equal numbers may be **added to** both sides of an equation.

2. *Subtraction Rule of Equality* — To maintain an equality, equal numbers may be **subtracted from** both sides of an equation.

3. *Multiplication Rule of Equality* — To maintain an equality, both sides of an equation may be **multiplied by** equal numbers.

4. *Division Rule of Equality* — To maintain an equality, both sides of an equation may be **divided by** equal numbers, except division by zero is undefined.

Problem 2.2
State the equality rule used to solve each problem:

    a) $x + 15 = 21$ b) $40 = r - 8$ c) $25 = 5m$

$$\begin{array}{c} \underline{-15 = -15} \\ x = 6 \end{array} \qquad \begin{array}{c} \underline{+8 = +8} \\ 48 = r \end{array} \qquad \begin{array}{c} \dfrac{25}{5} = \dfrac{5m}{5} \\ 5 = m \end{array}$$

Ans.   a) Subtraction Rule   b) Addition Rule   c) Division Rule

Using Addition to Solve an Equation

When using the addition rule of equality to solve an equation follow the procedure below:

Solve: $n - 19 = 21$

Procedure:	Solution:
1. Add to both sides the number subtracted from the unknown:	$\begin{array}{r} n - 19 = \quad 21 \\ \underline{+19 = +19} \\ n = \quad 40 \quad Ans. \end{array}$
2. Check the original equation:	$\begin{array}{r} n - 19 = \quad 21 \\ 40 - 19 =^? 21 \\ 21 = 21 \quad (Yes) \end{array}$

Example 2.4
Solve the equation:

    a) $12.5 = m - 2.9$

Ans.   a) 1. Add to both sides
            the number subtracted
            from the unknown:

$$\begin{array}{r} 12.5 = m - 2.9 \\ \underline{+2.9 = \quad +2.9} \\ 15.4 = m \quad Ans. \end{array}$$

2. Check the original equation: $12.5 = m - 2.9$
$12.5 =^? 15.4 - 2.9$
$12.5 = 12.5$ (Yes)

Problem 2.3
Solve:

a) After giving 15 marbles to Sam, Mario has 43 left. How many marbles did Mario have originally?

Ans. a) Mario had 58 marbles originally.

Using Subtraction to Solve an Equation

When using the subtraction rule of equality to solve an equation follow the procedure below:

Solve: $w + 12 = 19$

Procedure: Solution:
1. Subtract from both sides
the number added to the unknown:
$$w + 12 = 19$$
$$\underline{ - 12 = -12}$$
$$w = 7 \quad Ans.$$

2. Check the original equation:
$$w + 12 = 19$$
$$7 + 12 =^? 19$$
$$19 = 19 \quad (Yes)$$

Example 2.5
Solve the equation:

a) $20.8 = d + 6.9$

Ans. a) 1. Subtract from both sides
the number added to the unknown: $20.8 = d + 6.9$
$$\underline{-6.9 = -6.9}$$
$$13.9 = d \quad Ans.$$

2. Check the original equation:
$$20.8 = d + 6.9$$
$$20.8 =^? 13.9 + 6.9$$
$$20.8 = 20.8 \quad \text{(Yes)}$$

Problem 2.4
Solve:

a) Pam's height is 5 ft 3 in. If she is 9 in taller than John, how tall is John?

Ans. a) John is 4 ft 6 in tall.

Using Multiplication to Solve an Equation

When using the multiplication rule of equality to solve an equation follow the procedure below:

Solve: $\dfrac{w}{3} = 5$

Procedure: Solution:
1. Multiply both sides
   of the equation by the
   divisor of the unknown:

$$\frac{w}{3} = 5$$

$$3 \cdot \frac{w}{3} = 5 \cdot 3$$
$$w = 15 \quad Ans.$$

2. Check the original equation:

$$\frac{w}{3} = 5$$

$$\frac{15}{3} =^? 5$$

$$5 = 5$$

You Need to Know ✔

To divide by a fraction, invert the fraction and multiply.

Thus, $8 \div \dfrac{2}{3} = 8 \times \dfrac{3}{2} = 12$.

Example 2.6
Solve the equation:

a) $\dfrac{4}{3}w = 30$

Ans. a) 1. Invert and multiply: $\dfrac{4}{3}w = 30$

$$\dfrac{3}{4} \cdot \dfrac{4}{3}w = 30 \cdot \dfrac{3}{4}$$

$$w = 22.5 \quad Ans.$$

2. Check the original equation: $\dfrac{4}{3}w = 30$

$$\dfrac{4}{3}(22.5) =^? 30$$
$$30 = 30 \quad (\text{Yes})$$

Problem 2.5

a) After traveling 84 miles Henry found that he had gone three-fourths of the entire distance to home. What is the total distance to his home?

Ans. a) The total distance is 112 miles.

Using Division to Solve an Equation

When using the division rule of equality to solve an equation follow the procedure below:

Solve: $2n = 16$

<table>
<tr><td>Procedure:</td><td>Solution:</td></tr>
</table>

Procedure:
1. Divide both sides of the equation by the multiplier of the unknown:

Solution:

$2n = 16$

$$\frac{2n}{2} = \frac{16}{2}$$

$n = 8$ *Ans.*

2. Check the original equation.

$2n = 16$

$2(8) =^? 16$

$16 = 16$ (Yes)

Example 2.7
Solve the equation:

    a) $75\%t = 18$
        Remember to change % to a decimal.

Ans.   a) 1. Divide both sides of the equation by the multiplier of the unknown:

$75\%t = 18$

$$\frac{.75t}{.75} = \frac{18}{.75}$$

$t = 24$ *Ans.*

2. Check the original equation.

$75\%t = 18$

$(.75)(24) =^? 18$

$18 = 18$ (Yes)

Example 2.8
Solve:

a) Mr. Wang's commission rate was 6 percent. If he earned $66 in commission, how much did he sell?

Ans. a) 6% of x is $66
.06x = 66
Divide by .06.
So x = $1100.

Using Two or More Operations to Solve an Equation

In equations where two operations are performed upon the unknown, two inverse operations are needed to solve the equation. When solving for an equation of this type follow the procedure below:

Solve: $2x + 7 = 19$

Procedure:
1. Perform addition to undo subtraction, or subtraction to undo addition:

Solution:

$$2x + 7 = 19$$
$$\underline{-7 = -7}$$
$$2x = 12$$

2. Perform multiplication to undo division, or division to undo multiplication:

$$2x = 12$$

$$\frac{2x}{2} = \frac{12}{2}$$

$$x = 6 \quad Ans.$$

3. Check the original equation.

$$2x + 7 = 19$$
$$2(6) + 7 =^? 19$$
$$19 = 19 \quad (\text{Yes})$$

Remember

When solving equations using two or more inverse operations, first perform addition and subtraction, then perform multiplication and division. This is an undo process so it uses the reverse order of operations.

Example 2.9
Solve the equation:

a) $13n + 4 + n = 39$

Ans.   a)   1. Collect like terms. Perform addition to undo subtraction, or subtraction to undo multiplication:

$$13n + 4 + n = 39$$
$$\underline{-4 = -4}$$
$$14n = 35$$

2. Perform multiplication to undo division, or division to undo multiplication:

$$14n = 35$$

$$\frac{14n}{14} = \frac{35}{14}$$

$$n = 2.5 \quad Ans.$$

3. Check the original equation:

$$13n + 4 + n = 39$$
$$13(2.5) + 4 + 2.5 =^? 39$$
$$39 = 39 \quad (Yes)$$

Problem 2.6

  a) How many boys are there in a class of 36 pupils if the total number of girls is 6 more than the total number of boys?

Ans. a) There are 15 boys.

Solving Equations Containing Fractions

Fractions with the Same Denominator

To solve equations having the same denominator or only one denominator other than 1, simply clear the fraction from the equation. When solving for an equation of this type follow the procedure below:

Solve: $\dfrac{x}{3} + 5 = 2x$

Procedure:

1. Clear fractions by multiplying both sides of the equation by the denominator:

2. Solve the resulting equation:

Solution:

$$\frac{x}{3} + 5 = 2x$$

Multiply by denominator 3

$$3\left(\frac{x}{3} + 5\right) = 3(2x)$$

$x + 15 = 6x$

$3 = x$ *Ans.*

Example 2.10

Solve the equation:

  a) $\dfrac{3x}{7} - 2 = \dfrac{x}{7}$

Ans. a) 1. Clear fractions by multiplying both sides of the equation by the denominator:

$$\frac{3x}{7} - 2 = \frac{x}{7}$$

Multiply by denominator 7

$$7\left(\frac{3x}{7} - 2\right) = 7\left(\frac{x}{7}\right)$$

2. Solve the resulting equation:

$$7\left(\frac{3x}{7}\right) - 7(2) = 7\left(\frac{x}{7}\right)$$

$$3x - 14 = x$$

$$2x = 14$$

$$x = 7$$

Fractions with Different Denominators

To solve equations having different denominators, find the lowest common denominator (LCD) first. The LCD is the smallest number divisible by their denominators without a remainder. Thus, in the equation:

$$\frac{1}{2} + \frac{x}{3} = \frac{7}{4}$$

12 is the lowest common denominator, since 12 is the smallest number divisible by the denominators 2, 3, and 4 without a remainder.

After the lowest common denominator is found, to solve the equation, clear the fractions from the equation by multiplying both sides by the lowest common denominator. Oftentimes the LCD is just the product of the denominators. When solving for an equation with fractions that have different denominators follow the procedure outlined below:

Solve: $\dfrac{x}{2} + \dfrac{x}{3} = 20$

Procedure:	Solution:
1. Clear fractions by multiplying both sides of the equation by the LCD:	$\dfrac{x}{2} + \dfrac{x}{3} = 20$

Multiply by LCD of 6

$$6\left(\frac{x}{2} + \frac{x}{3}\right) = 6(20)$$

2. Solve the resulting equation:

$$3x + 2x = 120$$
$$5x = 120$$
$$x = 24 \ \textit{Ans.}$$

Example 2.11
Solve the equation:

    a) $\dfrac{a}{2} - \dfrac{a}{3} - \dfrac{a}{5} = 2$

Ans.   a)   1. Clear fractions by multiplying both sides of the equation by the denominator:

$$\frac{a}{2} - \frac{a}{3} - \frac{a}{5} = 2$$

Multiply by LCD of 30

$$30\left(\frac{a}{2} - \frac{a}{3} - \frac{a}{5}\right) = 30(2)$$

    2. Solve the resulting equation:

$$15a - 10a - 6a = 60$$
$$a = -60 \ \textit{Ans.}$$

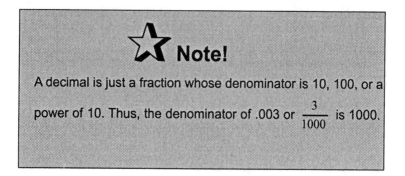

To solve equations using decimals, mulitply both sides of the equation by the denominator of the decimal having the greatest number of decimal places. Use the procedure below to solve these types of equations:

Solve: $8 = .05b$

Procedure: Solution:
1. Clear decimals by
    multiplying both sides
    of the equation by the
    denominator having the
    greatest number of decimal places: $8 = .05b$

$$8 = \frac{5}{100}b$$

Multiply by 100

2. Solve the resulting equation: $800 = 5b$
                                                    $160 = b$

Solving Literal Equations

Literal equations contain two or more letters. Thus $x + y = 20$, $5x = 15a$, and $D = RT$ are all literal equations. To solve a literal equation for any letter, follow the same procedure used in solving any equation for an unknown. Thus to solve for x in $5x = 25a$, divide both sides by 5 to obtain $x = 5a$.

Example 2.12
Solve for y.

  a) $2y - 4a = 8a$ b) $3(y - 2b) = 9a - 15b$

Ans.   a) $2y = 8a + 4a$ b) $3y - 6b = 9a - 15b$
          $2y = 12a$ $3y = 9a - 9b$
          $y = 6a$ *Ans.* $y = 3a - 3b$ *Ans.*

Chapter 3
GRAPHS OF LINEAR EQUATIONS

Understanding Graphs

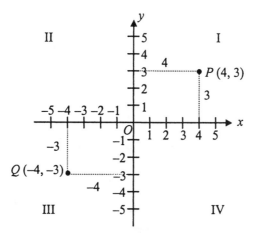

Figure 3-1

The origin O is the point where the two axes or number lines cross each other. To locate a point, determine its distance from each axis. Note on the graph that for P, the distances are +4 and +3. The coordinates of a point are its distances from the axes, with the appropriate signs attached.

1. **The x coordinate of a point**, its **abscissa**, is its distance from the y axis. This distance is measured along the x axis. For P, this is +4 ; for Q, this is –4.

2. **The y coordinate of a point**, its **ordinate**, is its distance from the x axis. This distance is measured along the y axis. For P, this is +3 ; for Q, this is –3.

3. In stating the coordinates of a point, the x coordinate precedes the y coordinate, just as in the alphabet x precedes y. Place the coordinates in parentheses. Thus, the coordinates of P are written (+4,+3) or (4,3), those for Q, are written (–4,–3).

The **quadrants of a graph** are the four parts cut off by the axes. Note on the graph how these are numbered I, II, III, and IV in a counter-clockwise direction.

Problem 3.1

On the graph shown in Figure 3-2, locate each point.

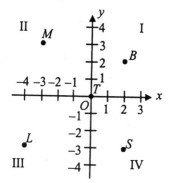

Figure 3-2

Point	Ans.	Coordinates
a) B		a) $(+2,+2)$ or $(2,2)$
b) M		b) $(-3,+3)$ or $(-3,3)$
c) T		c) $(0,0)$, the origin
d) L		d) $(-4,-3)$
e) S		e) $(+2,-3)$ or $(2,-3)$

Example 3.1

a) On the graph shown in Figure 3-3, if $A,B,C,$ and D are the vertices of the rectangle shown, find its perimeter and area.

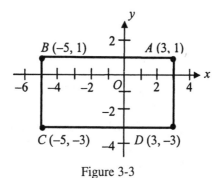

Figure 3-3

Ans. a) The base and the height of rectangle *ABCD* are 8 and 4 respectively. Hence, the perimeter is 24 inches, 2(8) + 2(4), and the area is 32, 8 × 4, square units.

Graphing Linear Equations

A **linear equation** is an equation whose graph is a straight line. To graph a linear equation, follow the procedure outlined below:

Graph: $y = x + 4$

Procedure:	Solution:
1. Make a table of coordinates for three pairs of values as follows: Let *x* have convenient values such as 2, 0, and –2. Substitute each of these for *x*, and find the corresponding value of *y*.	Table of coordinate values: Since $y = x + 4$, If $x = 2$, $y = 2 + 4 = 6$, so A = (2,6) If $x = 0$, $y = 0 + 4 = 4$, so B = (0,4) If $x = -2$, $y = -2 + 4 = 2$, so C = (-2,2)
2. Plot the points and draw the straight line joining them:	See Figure 3-4 *Ans.*

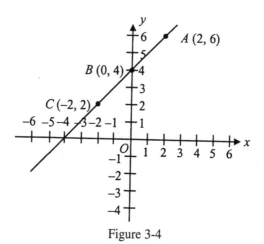

Figure 3-4

Two points determine a line. A third point is recommended in Step 1 in the procedure outlined above to serve as a checkpoint to ensure correctness.

An **intercept of a graph** is the distance from the origin to the point where the graph crosses either axis.

1. The **x intercept** of a graph is the value of x for the point where the graph crosses the x-axis. At this point, $y = 0$. In Figure 3-4, the x intercept is –4.

2. The **y intercept** of a graph is the value of for the point where the graph crosses the y-axis. At this point, $x = 0$. In Figure 3-4, the y intercept is 4.

Equations of the First Degree

1. **An equation of the first degree in one unknown** is one, which after it has been simplified, contains only one unknown having the exponent 1. Thus, $2x = 7$ is an equation of the first degree in one unknown. The unknown is called a **variable.**

2. **An equation of the first degree in two unknowns** is one which, after it has been simplified, contains only two unknowns, each of them in a separate term and having the exponent 1. Thus, $2x = y + 7$ is an equation of the first degree in two unknowns, but $2xy = 7$ is not since x and y are not in separate terms.

• *Rule 1* — The graph of a first-degree equation in one or two unknowns is a straight line.

• *Rule 2* — The graph of a first-degree equation in only one unknown is the x-axis, the y-axis, or a line parallel to one of the axes.

Important

The graph of $y = 0$ is the x-axis and the graph of $x = 0$ is the y-axis.

• *Rule 3* — If a point is on the graph of an equation, its coordinates satisfy the equation.

• *Rule 4* — If a point is not on the graph of an equation, its coordinates do not satisfy the equation.

Solving a Pair of Linear Equations Graphically

The **common solution** of two linear equations is the **one and only one** pair of values that satisfies both equations. Thus, $x = 3$, $y = 7$ is the common solution of the equations $x + y = 10$ and $y = x + 4$. Note that the graphs of these equations meet at the point $x = 3$ and $y = 7$. Since two intersecting lines meet in one and only one point, this pair of values is the common solution.

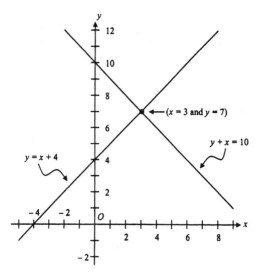

Figure 3-5

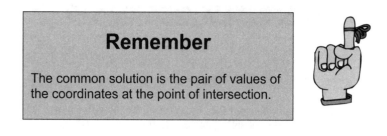

Remember

The common solution is the pair of values of the coordinates at the point of intersection.

Consistent, Inconsistent, and Dependent Equations

1. Equations are **consistent** if one and only one pair of values satisfies both equations. Note: Two distinct intersecting lines. See Figure 3-5.

2. Equations are **inconsistent** if no pairs of values satisfy both equations. Note: Two parallel lines. See Figure 3-6.

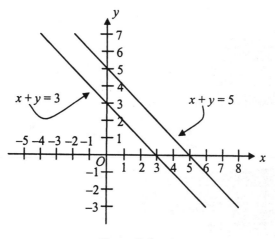

Figure 3-6

3. Equations are **dependent** if any pair of values that satisfies one also satisfies the other. **Note:** two equations for the same line obtained by performing the same operation with the same number to both sides. See Figure 3-7.

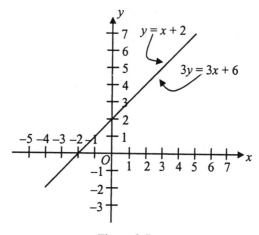

Figure 3-7

Problem 3.2

From the graphs of $3y + x = 5$, $x + y = 3$, and $y = x + 3$ shown in Figure 3-8, find the common solution of:

    a) $3y + x = 5$ and $x + y = 3$
    b) $3y + x = 5$ and $y = x + 3$
    c) $x + y = 3$ and $y = x + 3$

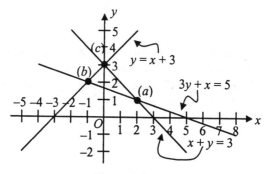

Figure 3-8

Ans. a) (2, 1)
  b) (–1,2)
  c) (0,3)

Solving a Pair of Equations by Addition or Subtraction

To solve a pair of equations by adding or subtracting follow the procedure outlined below:

Solve: $3x - 7 = y$
   $4x - 5y = 2$

Procedure:	Solution:
1. Arrange so that like terms are in the same column:	$3x - y = 7$ $4x - 5y = 2$
2. Multiply so that the coefficient of one of the unknowns will have the same absolute value:	Multiply $3x - y = 7$ by 5 $15x - 5y = 35$
3. To eliminate the unknown whose coefficients have the same absolute value: a) Add if the signs are unlike b) Subtract if the signs are like	Eliminate y $15x - 5y = 35$ $- (4x - 5y = 2)$ ———————— $11x \quad = 33$
4. Solve the resulting equation:	$x = 3$

5. Find the other unknown
    by substituting the value
    found in any equation having
    both unknowns:

    Find y:
    $$4x - 5y = 2$$
    $$4(3) - 5y = 2$$
    $$y = 2$$
    $x = 3$ and $y = 2$ or $(3,2)$ *Ans.*

Note: You may have to multiply both equations by different numbers
to eliminate a variable.

Problem 3.3
Use addition or subtraction to eliminate one unknown and solve for x
and y.

a) $5x + 3y = 19$ *Ans.*   a) $(2, 3)$
  $x + 3y = 11$

b) $3x - 5y = 19$ b) $(-2, -5)$
  $2x - 4y = 16$

Solving a Pair of Equations by Substitution

To solve a pair of equations by substitution follow the procedure outlined below:

Solve: $x - 2y = 7$
   $3x + y = 35$

Procedure: Solution:
1. Solve the equation for   Express x in terms of y:
    one of the variables: $x = 2y + 7$

2. Substitute this value for
   the variable in the second equation: Substitute for x in
   $$3x + y = 35:$$
   $$3(2y + 7) + y = 35$$

3. Find the remaining unknown: $$6y + 21 + y = 35$$
   $$7y = 14$$

4. Solve the resulting equation: $$y = 2$$

5. Solve for the other variable
   by substituting into one of the
   original equations: Substitute 2 for y :
   $$x = 2(2) + 7$$
   $$x = 11$$
   so $x = 11$ and $y = 2$ *Ans.*

Problem 3.4
Solve by substitution:

a) $x - y = 12$ *Ans.*   a) $(8, -4)$
   $3x = x - 4y$

b) $x = 2(y - 5)$ b) $(-12, -1)$
   $4x + 40 = y - 7$

Deriving a Linear Equation from a Table of Values

Deriving a Simple Linear Equation by Inspection

A simple linear equation such as $y = x + 3$ involves only one operation. Any y value equals 3 **added to** its corresponding x value. Note this in the following table of values for $y = x + 3$:

y	−7	0	3	4	6	13
x	−10	−3	0	1	3	10

Deriving a Simple Linear Equation by the Ratio Method

How can we derive a linear equation when two operations are involved, as in the case of $y = 4x + 2$? In the table below, any y value equals 2 added to 4 times its corresponding x value. Examine the following table of values for $y = 4x + 2$ and notice that as x increases 1, y increases 4; as x increases 2, y increases 8; and as x decreases 1, y decreases 4.

Change in y +4   +8   −4

y	6	10	18	14
x	1	2	4	3

Change in x +1   +2   −1

Compare each change in y (above the table) with the corresponding change in x (below the table). Notice that the y difference equals 4 times the x difference. From this, we may conclude that the equation is of the form $y = 4x + b$. The value of b may now be found. To find b, substitute any pair of values for x and y in $y = 4x + b$.

> Thus, in $y = 4x + b$, substitute $x = 1$ and $y = 6$.
> $6 = 4(1) + b$
> $2 = b$ Since $b = 2$, the equation is $y = 4x + 2$.

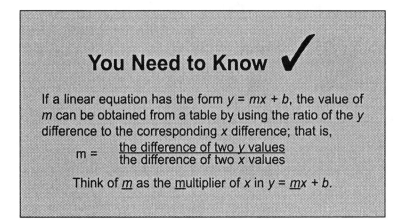

Midpoint of a Segment

The coordinates (x_m, y_m) of midpoint M of the line segment joining $P(x_1, y_1)$ and $Q(x_2, y_2)$ are:

$$x_m = \frac{1}{2}(x_1 + x_2) \text{ and } y_m = \frac{1}{2}(y_1 + y_2)$$

Example 3.2

   a) If M is the midpoint of segment PQ, find the coordinates of M if the coordinates of P are $P(3,4)$ and coordinates of Q are $Q(5,8)$.

Ans. a) Average the x coordinates: $3 + 5 = 8$, $8/2 = 4$
Average the y coordinates: $4 + 8 = 12$, $12/2 = 6$
$(4,6)$ *Ans.*

Distance between Two Points

• *Rule 1* — The distance between two points having the same ordinate (or y value) is the absolute value of the difference of their abscissas (or x value). Hence, the distance between two points must be positive. Thus, the distance between $P(6,1)$ and $Q(9,1)$ is $9 - 6 = 3$.

• *Rule 2* — The distance between two points having the same abscissa (or x value) is the absolute value of the difference of their ordinates (or y value). Thus, the distance between $P(2,1)$ and $Q(2,4)$ is $4 - 1 = 3$.

• *Rule 3* — If the points do not have the same x or y value, the distance between points $P(x_1, y_1)$ and $Q(x_2, y_2)$ is:

$$D = \sqrt{(x_2 - x_1)^2 + (y_2 - y_1)^2}$$

Thus, the distance between $P(2,1)$ and $Q(5,5)$ is:

$$D = \sqrt{(5-2)^2 + (5-1)^2} = \sqrt{9+16} = \sqrt{25} = 5$$

Don't Forget!

$$\sqrt{9+16} \neq 3 + 4 \neq 7$$

Problem 3.5
Find the distance between each of the following pairs of points:

a) $(-3, -6)$ and $(3, 2)$ *Ans.*   a) 10

b) $(2, 2)$ and $(5, 5)$ b) $\sqrt{18} = 3\sqrt{2}$

Chapter 4
MONOMIALS
AND
POLYNOMIALS

IN THIS CHAPTER:

✔ Understanding Monomials
  and Polynomials
✔ Adding and Subtracting Monomials
  and Polynomials
✔ Multiplying Monomials
  and Polynomials
✔ Dividing Monomials and Polynomials

Understanding Monomials and Polynomials

A **term** is a number or the product of numbers. Each of these numbers is a **factor** of the term. Terms are numbers that are added or subtracted. Factors are numbers that are multiplied. The term $-5xy$ consists of three factors -5, x, and y. In $-5xy$, xy is the variable coefficient and -5 is the numerical coefficient.

57

An **expression** consists of one or more terms.

1. A **monomial** is an expression of one term.

2. A **polynomial** is an expression of two or more terms.

   a) A **binomial** is a polynomial of two terms.
      $2x - 3$ is a binomial.

   b) A **trinomial** is a polynomial of three terms.
      $3x^2 - 2x + 1$ is a trinomial.

Like terms are terms having the same variable factors, each with the same exponent. $2xy^2$ and $-5xy^2$ are like terms. **Unlike terms** are terms that do not have the same variable coefficients. $2xy$ and $3x^2$ are unlike terms.

Adding and Subtracting Monomials and Polynomials

Adding and Subtracting Monomials

To add or subtract like terms:

1. Add or subtract their numerical coefficients.

2. Keep the common variable coefficient.

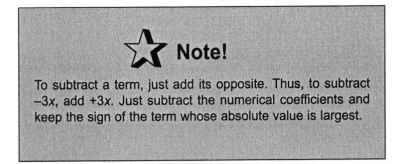

☆ Note!

To subtract a term, just add its opposite. Thus, to subtract $-3x$, add $+3x$. Just subtract the numerical coefficients and keep the sign of the term whose absolute value is largest.

Example 4.1

Simplify the following:

a) $+5a + (-2a) - (-4a)$ *Ans.* a) $5a - 2a + 4a$
$$= 7a$$

b) $+10x + (-5a) - (+2x) + (+3a)$ *Ans.* b) $10x - 5a - 2x + 3a$
$$10x - 2x - 5a + 3a$$
$$= 8x - 2a$$

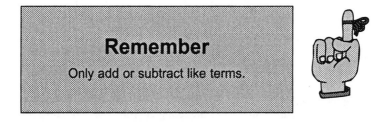

Remember

Only add or subtract like terms.

Arranging and Adding Polynomials

A polynomial may be arranged as follows:

1. In **descending order,** by having the exponents of the same variable decrease in successive terms. $-x^4 + x^3 + x^2 + x$

2. In **ascending order,** by having the exponents of the same variable increase in successive terms. $-x + x^2 + x^3 + x^4$

Adding and Subtracting Polynomials

To add or subtract polynomials:

1. Arrange polynomials in order, placing like terms in the same column.

2. Add or subtract like terms.

With polynomials the sign change must be made for all elements of the equation. Thus, to subtract $3x + 2y$, add $-3x - 2y$.

Example 4.2
Add the following polynomials:

a) $6x - 2y + 4$ and $4y - 5 - 2x$

Ans. a) <u>Procedure:</u>   <u>Solution:</u>

      1. Arrange polynomials $6x - 2y + 4$
        in order, placing like $+ \underline{\ -2x + 4y - 5}$
        terms in the same column:

      2. Add like terms: $4x + 2y - 1$ *Ans.*

Parentheses and other grouping symbols such as brackets [ ] and braces { } are used in adding and subtracting polynomials. The rules for removing these grouping symbols are:

- *Rule 1* — When removing parentheses preceded by a plus sign, **do not change** the signs of the enclosed terms.
  Thus, $3x + (+5x - 10) = 3x + 5x - 10$.

- *Rule 2* — When removing parentheses preceded by a minus, **change** the sign of each enclosed term.
  Thus, $3x - (+5x - 10) = 3x - 5x + 10$

- *Rule 3* — When more than one set of grouping symbols is used, remove one set at a time, beginning with the innermost one. Thus, $2 + [r - (3 - r)]$
  $$2 + [r - 3 + r]$$
  $$2 + r - 3 + r$$
  $$2r - 1$$

Example 4.3
Subtract the following polynomials:

a) $(-2x + 4y - 5) - (6x - 2y + 4)$

Ans. a) <u>Procedure:</u>   <u>Solution:</u>
   1. Arrange polynomials
      in order, placing like
      terms in the same column: $-2x + 4y - 5$
                                  $-\ \ \underline{(6x - 2y + 4)}$

   2. Change each sign in
      the subtracted polynomial: $-2x + 4y - 5$
                                  $+\ \ \underline{-6x + 2y - 4}$

   3. Add like terms: $-8x + 6y - 9$ *Ans.*

Multiplying Monomials and Polynomials

Multiplying Monomials and Numbers with the Same Base

• *Rule 1* — To multiply numbers with the same base, keep the base
and add the exponents. Thus, $(x^4)(x^5) = x^9$. Why?

$x^4 = x \cdot x \cdot x \cdot x$ and $x^5 = x \cdot x \cdot x \cdot x \cdot x$ so $(x^4)(x^5) =$

$x \cdot x \cdot x \cdot x \cdot x \cdot x \cdot x \cdot x \cdot x = x^9$

• *Rule 2* — To find the power of a power, keep the base and multiply
the exponents. Thus, $(x^4)^5 = x^{20}$. Why?

$(x^4)^5 = x^4 \cdot x^4 \cdot x^4 \cdot x^4 \cdot x^4 = x^{20}$

- *Rule 3* — Changing the order of factors does not change their product. This fundamental law is known as the commutative law of multiplication. Thus, $3x(4x^2)= 3(4)(x)(x^2) = 12x^3$

Multiplying Monomials

To multiply monomials by each other follow the procedure outlined below:

Multiply: $2x \cdot 3x^2 \cdot -4x^3$

Procedure:
1. Multiply numerical coefficients:
2. Multiply variables:
3. Multiply results:

Solution:

$(2)(3)(-4) = -24$
$(x)(x^2)(x^3) = x^6$
$= -24x^6$ *Ans.*

Problem 4.1
Multiply.

  a) $(2x^2y^3)(-3x^3)(-4xy)$

Ans.   a) $24x^6y^4$

Multiplying a Polynomial by a Monomial

- *Rule 1* — To multiply a polynomial by a monomial, multiply each term of the polynomial by the monomial. Thus, $4(a + b)$ $= 4a + 4b$. This fundamental law is known as the **distributive law.**

Problem 4.2
Multiply:

  a) $3(x - 2)$ *Ans.*   a) $3x - 6$

b) $x(3x + 1)$

c) $2(x - y) + 3(x + 2y)$

b) $3x^2 + x$

c) $2x - 2y + 3x + 6y$
$= 5x + 4y$

Multiplying Polynomials

To multiply polynomials by each other follow the procedure outlined below:

Multiply: $(3x + 4)(1 + 2x)$

<u>Procedure:</u>   <u>Solution:</u>

1. Arrange each
   polynomial in order: $(3x + 4)(2x + 1)$
2. Multiply each term of
   one polynomial by each
   term of the other polynomial: $6x^2 + 3x + 8x + 4$
3. Add like terms: $6x^2 + 11x + 4$

Multiply using **FOIL** – multiply the First terms, Outside terms, Inside terms, Last terms. Then add like terms.

Problem 4.3
Multiply:

> a) $(8 + c)(3 - 2c)$

Ans. a) $-2c^2 - 13c + 24$

Dividing Monomials and Polynomials

Rules for Dividing Monomials

When dividing powers that have the same base, arrange them into a fraction and apply the following rules:

- *Rule 1* — If the exponent of the numerator is larger than the exponent of the denominator, keep the base and subtract the smaller exponent from the larger. Thus, $\dfrac{x^a}{x^b} = x^{a-b}$.

- *Rule 2* — If the exponents are equal, then we have a number divided by itself; the quotient is 1. Thus, $\dfrac{x^4}{x^4} = 1$.

- *Rule 3* — If the exponent of the denominator is larger, make the numerator of the quotient 1, and to obtain its denominator, keep the base and subtract the smaller exponent from the larger. Thus, $\dfrac{x^3}{x^7} = \dfrac{1}{x^{7-3}} = \dfrac{1}{x^4}$.

Problem 4.4
Divide:

a) $\dfrac{-7ab^3c}{14a^4b^2c}$

Ans. a) $\dfrac{-b}{2a^3}$

Dividing a Polynomial by a Monomial

To divide a polynomial by a monomial, divide each term of the polynomial by the monomial.

Thus, $\dfrac{10x+15}{5} = \dfrac{10x}{5} + \dfrac{15}{5} = 2x+3$.

Problem 4.5
Divide:

a) $\dfrac{9x^2 - 36xy^2}{9xy}$

Ans. a) $\dfrac{x}{y} - 4y$

Dividing a Polynomial by a Polynomial

To divide polynomials by each other follow the procedure outlined below:

Divide: $(x^2 - 5x + 6)$ by $(x - 2)$

Procedure: Solution:

1. Set up as a form
   of long division
   putting each polynomial $x-2\overline{)x^2-5x+6}$
   in descending order:

2. Divide the first term
   of the divisor in the $x-2\overline{)x^2-5x+6}^{\quad\ x}$
   first term of the dividend:

3. Multiply the first term of
   the quotient by each term $x-2\overline{)x^2-5x+6}^{\quad\ x}$
   of the divisor:
                                                    $\underline{x^2-2x}$

4. Subtract like terms and
   bring down one or more
   terms as needed: $-3x+6$
5. Repeat steps 2–4:
   Divide -3
   Multiply $x-2\overline{)-3x+6}$
   Subtract $\underline{-3x+6}$

6. If there is a remainder, it
   is added as a fraction of
   the divisor: 0

                                                    $x-3$ Ans.

Example 4.4
Divide:

a) $(8x^2 - 10x + 8)$ by $(2x - 4)$

Ans. a)

$$2x-4 \overline{)\begin{array}{r} 4x+3+\dfrac{20}{2x-4} \\ 8x^2 -10x+8 \end{array}}$$

$$\underline{8x^2 - 16x}$$
$$6x + 8$$
$$\underline{6x - 12}$$
$$20$$

Chapter 5
PROBLEM
SOLVING

Introduction to Problem Solving

The four steps of problem solving are as follows:

1. **Representation** of unknowns by variables (preceded, obviously, by identification of all unknowns)!

2. **Translation** of relationships about unknowns into equations

3. **Solution** of equations to find the values of the unknowns

4. **Check** values found to see if they satisfy the original problem.

Consecutive–Integer Problems

Each consecutive–integer problem involves a set of consecutive integers, a set of consecutive even integers, or a set of consecutive odd integers. Each such set involves integers arranged in increasing order from left to right.

NOTE: In the table in Figure 5-1, n represents the first number of a set. However, n may be used to represent any other number in the set. Thus, a set of three consecutive integers may be represented by $n - 1$, n, and $n + 1$.

Table of Integers

	Consecutive Integers	Consecutive Even Integers	Consecutive Odd Integers
Illustrations	4, 5, 6, 7 −4, −3, −2, −1	4, 6, 8, 10 −4, −2, 0, 2	5, 7, 9, 11 −5, −3, −1, 1
Kinds of Integers	Odd or even	Even only	Odd only
Differ by	1	2	2
Representation of first consecutive number of second consecutive number of third consecutive number	n $n + 1$ $n + 2$	n $n + 2$ $n + 4$	n $n + 2$ $n + 4$

Figure 5-1

Example 5.1

a) Find five consecutive odd integers whose sum is 45.

Ans. a) *Method I*

1. Represent the five consecutive odd integers using n, $n + 2$, $n + 4$, $n + 6$, $n + 8$.
2. Therefore, their sum is $5n + 20$.
3. Thus, $5n + 20 = 45$ and $n = 5$. (the first number).
4. 5, 7, 9, 11, 13 *Ans.*

Method II

1. Represent the five consecutive odd integers using $n - 4$, $n - 2$, n, $n + 2$, $n + 4$.
2. Therefore, their sum is $5n$.
3. Then, $5n = 45$ and $n = 9$ (the third)
4. 5, 7, 9, 11, 13 *Ans.*

Age Problems

• *Rule 1 —* To find a person's future age in a number of years, **add** that number to her or his present age. Thus, in 10 years, a person 17 years old will be $17 + 10$ or 27 years old.

• *Rule 2 —* To find a person's past age a number of years ago, **subtract** that number of years from his or her present age. Thus, 10 years ago, a person 17 years old was $17 - 10$ or 7 years old.

Example 5.2
A father is now 20 years older than his daughter. In 8 years, the father's age will be 5 years more than twice the daughter's age then. Find their present age.

Ans. *Method I*

1. Let F = father's present age and let D = daughter's present age
2. Therefore, $F = D + 20$
3. Then, $F + 8 = 2(D + 8) + 5$
4. Substitute $D + 20$ for F

5. Therefore, $(D + 20) + 8 = 2(D + 8) + 5$
$$D + 28 = 2D + 21$$
$$D = 7 \quad Ans.$$

Method II

1. Let D = daughter's present age and $D + 20$ = father's present age

2. Therefore $(D + 20) + 8 = 2(D + 8) + 5$
$$D + 28 = 2D + 16 + 5$$
$$D = 7 \quad Ans.$$

Ratio Problems

Ratios are used to compare quantities by division. Ratios are relationships. A ratio can be expressed in the following ways:

1. Using a colon: 3:4

2. Using *to*:   3 to 4

3. As a common fraction: 3/4

4. As a decimal: 0.75

5. As a percent: 75%

General Principles of Ratios

• *Principle 1* — To find the ratios between quantities, the quantities must have the same unit. Thus, 1 foot to 4 inches is 12 inches to 4 inches or 3:1.

• *Principle 2* — A ratio is an abstract number, that is a number without a unit of measure. Thus, $12 to $4 is 3:1. The dollar sign is removed.

• *Principle 3* — A ratio should be simplified by reducing to lowest terms and eliminating fractions contained in the ratio. Thus, the ratio of 20 to 30 is 2:3.

• *Principle 4* — The ratio of three or more quantities may be expressed as a continued ratio. This is simply an enlarged ratio statement. Thus, the ratio of $2 to $3 to $5 is the ratio 2:3:5.

Problem 5.1

Express each ratio in lowest terms:

a) 175 to 75
b) $2x^3 : 6x$

Ans. a) 7:3
b) $x^2:3$

Example 5.3

a) Two numbers have a ratio of 2:5. If their sum is 35, what is the larger number?

Ans. a) Let $5x$ represent the larger number
1. Therefore, $2x + 5x = 35$
$$x = 5$$
2. Thus, the larger number, $5x$, equals 25. *Ans.*

Angle Problems

Pairs of Angles

1. **Adjacent angles** are two angles having the same vertex and a common side between them. In Figure 5-2, $\angle a$ and $\angle b$ are adjacent angles.

Figure 5-2

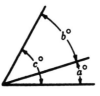

2. **Complementary angles** are two angles the sum of whose measures equals $90°$ or a right angle. In Figure 5-3a and Figure 5-3b, $\angle a$ and $\angle b$ are complementary angles.

Figure 5-3a   Figure 5-3b

3. **Supplementary angles** are two angles the sum of whose measures equals $180°$ or a straight angle. In Figure 5-4a and Figure 5-4b, $\angle a$ and $\angle b$ are supplementary angles.

Figure 5-4a   Figure 5-4b

4. The sum of the measures of the angles of any triangle equals $180°$. In Figure 5-5, $a + b + c = 180°$.

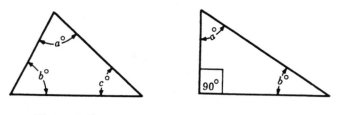

Figure 5-5   Figure 5-6

5. The sum of the measures of the acute angles of a right triangle equals 90°. In Figure 5-6, $a + b = 90°$.

6. The angles opposite the congruent sides of an isosceles triangle are congruent. In Figure 5.7, since AB = BC = 4x, then $\angle B \cong \angle C$.

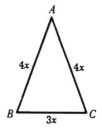

Figure 5-7

Example 5.4

    a) Two adjacent angles are in the ratio of 3 to 2 and form an angle of 40°. Find the angles.

Ans. a) 1. Let the angles = 3x and 2x, such that $3x + 2x = 40$.
        2. Then, solve for $x = 8$.

3. Then, the angles are 3(8) and 2(8) or $24°$ and $16°$. *Ans.*

Example 5.5

a) An angle is twice its complement. What is the angle?

Ans. a) 1. Let x = the complement of the angle and $2x$ = the angle.
   2. Therefore, $x + 2x = 90$, and $x = 30$.

   3. Thus, the angle is 2(30) or $60°$. *Ans.*

Example 5.6

a) An angle is $60°$ less than twice its supplement. Find the angles.

Ans. a) 1. Let x = the angle and $180 - x$ = the supplement of the angle.
   2. Therefore, $x = 2(180–x) – 60$
                     $x = 360 – 2x – 60$
                     $x = 100$

   3. Thus, the angles are $100°$ and $80°$. *Ans.*

Perimeter Problems

The perimeter of a polygon is the sum of its sides.

Thus, for a quadrilateral as shown below in Figure 5-8, if p = the perimeter, then:

$$p = a + b + c + d$$

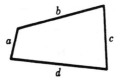

Figure 5-8

The perimeter of a regular polygon equals the product of one side and the number of sides. Thus for the regular pentagon shown in Figure 5-9, if p = the perimeter, then:

$$p = 5s$$

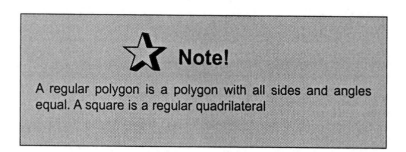

⭐ **Note!**

A regular polygon is a polygon with all sides and angles equal. A square is a regular quadrilateral

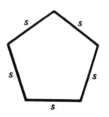

Figure 5-9

Example 5.7

    a) If the perimeter of the trapezoid shown in the figure below is 34, what are the lengths of the sides?

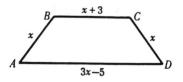

Figure 5-10

Ans. a) 1. Add the sides: $x + (x + 3) + x + (3x - 5) = 34$
2. Therefore $6x - 2 = 34$
$$6x = 36$$
$$x = 6$$
3. Thus, the lengths of the sides are 6, 9, 6, 13. *Ans.*
$$x = 6$$
$$x + 3 = 6 + 3 = 9$$
$$3x - 5 = 18 - 5 = 13$$

Coin Problems

The total value T of a number of coins of the same kind equals the number N of the coins multiplied by the value V of one of the coins.

$$T = NV$$

Thus, 8 nickels have total value of $8(5) = 40$ cents.

Example 5.8
a) In her coin bank, Maria has three times as many quarters as nickels. If the value of the quarters is \$5.60 more than the value of the nickels, how many of each kind does she have?

Ans. a) 1. Let n = number of nickels and $3n$ = number of quarters.
2. Let $5n$ = total value of nickels and $25(3n)$ = total value of quarters.
3. Therefore, $75n = 560 + 5n$.
4. Thus, $n = 8$. 8 nickels, 24 quarters. *Ans.*

Interest Problems

Annual interest I equals the principal P multiplied by the rate of interest R per year.

$$I = PR$$

Thus, annual interest from $200 at 6% per year is (200)(.06) or $12.

Example 5.9

   a) Mr. Wong invested $8,000, part at 5% and the rest at 2%. The yearly income from the 5% investment exceeded that from the 2% by $85. Find the investment at each rate.

Ans. a) 1. Let x = amount invested at 5% and $8000 - x$ = the amount invested at 2%.
   2. Therefore, A (Annual Interest for Part 1) = $.05x$ and
              B (Annual interest for Part II) = $.02(8000-x)$
   3. Thus, $A - B = .05x - .02(8000 - x) = \$85$.
   4. Multiply both sides by 100.
   5. Therefore, $5x - 2(8,000 - x) = 8500$
              $5x - 16,000 + 2x = 8500$
                        $7x = 24,500$
   6. Thus, $x = \$3,500$ and $8000 - x = \$4,500$. *Ans.*

Motion Problems

The distance D traveled equals the rate or speed R multiplied by the time spent traveling T.

$$D = RT \qquad\qquad R = \frac{D}{T} \qquad\qquad T = \frac{D}{R}$$

Thus, the distance traveled in 5 hours at a rate of 30 miles per hour is 150 miles.

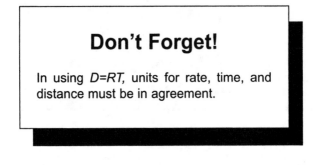

Don't Forget!

In using *D=RT,* units for rate, time, and distance must be in agreement.

Example 5.10

a) In going 1860 miles, William used a train first and later a plane. The train, going at 30 miles per hour took 2 hours longer than the plane traveling at 150 miles per hour. How long did the trip take?

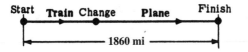

Start Train Change Plane Finish

|← 1860 mi →|

Figure 5-11

Ans. a)

	Rate (mph)	Time (h)	Distance (mi)
Plane trip	150	T	$150T$
Train trip	30	$T+2$	$30T+60$

1. The total distance is 1860 miles.
2. Therefore, $150T + 30T + 60 = 1860$.
3. Solve for T. $T = 10$
4. The trip took 22 hours (10 by plane and 12 by train). *Ans.*

Chapter 6
FACTORING

Understanding Factors and Products

A **product** is the result obtained by multiplying two or more numbers. The **factors of the product** are the numbers being multiplied. Thus, 2, x, and y are the factors of the product $2xy$. To find the **product of a monomial and a polynomial,** multiply the monomial by every term of the polynomial.

Thus, $3(4x + 2y) = 12x + 6y$

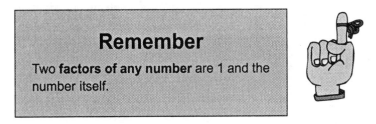

Remember

Two **factors of any number** are 1 and the number itself.

To **factor a number or expression** is to find its factors, not including 1 and itself. Thus, $2ax + 6a$ may be forced into $2(ax + 3a)$ or further into $2a(x + 3)$. To **factor a polynomial (two or more terms) completely,** continue factoring until the polynomial factors cannot be factored further. Thus, $2ax + 6a$ may be factored completely into $2a(x + 3)$.

Problem 6.1
Find the product:

    a) $-7x(x - 2)$

Ans.   a) $-7x^2 + 14x$

Problem 6.2
Factor completely:

    a) $4x^3 + 8x^2 - 24x$

Ans. a) $4x(x^2 + 2x - 6)$

Factoring a Polynomial Having a Common Monomial Factor

A **common monomial factor of a polynomial** is a monomial that is a factor of each term of the polynomial. Thus, 7 and a^2 are common monomial factors of $7a^2 + 7a^2y - 7a^2z$.

The **highest common monomial factor of a polynomial (HCF)** is the product of all its common monomial factors. Thus $7a^2$ is the highest common factor of the polynomial $7a^2 + 7a^2y - 7a^2z$.

Factor: $7ax^2 + 14bx^2 - 21cx^2$

Procedure:	Solution:
1. Use the HCF as one factor:	HCF is $7x^2$
2. Divide each term by the HCF to determine the other factor:	$\dfrac{7ax^2 + 14bx^2 - 21cx^2}{7x^2}$ $= a + 2b - 3c$
3. Display the results:	$7x^2(a + 2b - 3c)$ *Ans.*

Squaring a Monomial

The **square of a number** is the product of the number multiplied by itself. The number is used twice as a factor. Thus, the square of 7 is 49 or $(7)(7) = 49$.

Opposites have the same square. Thus, the square of –7 is also 49 or (–7)(–7) =49.

To **square a monomial,** square its numerical coefficient, keep each base, and double the exponent of each base. Thus, $(5ab^3)(5ab^3)$ $= 25a^2b^6$.

Finding the Square Root of a Monomial

The **square root of a number** is one of its two equal factors. Thus, the square root of 49 is either +7 or –7, since 49 = (7)(7) = (–7)(–7).

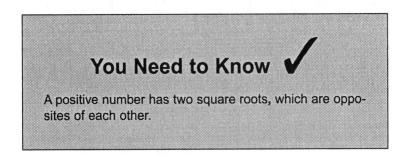

You Need to Know ✔

A positive number has two square roots, which are opposites of each other.

The **principal square root of a number** is its positive square root. **To find the principal square root of a monomial** find the principal square root of its numerical coefficient, keep each base, and use half the exponent of each base. Thus $\sqrt{16y^{16}} = 4y^8$.

Problem 6.3
Find the principal square root:

a) $\sqrt{\dfrac{x^{18}}{y^2}}$

Ans. a) $\dfrac{x^9}{y}$

Factoring the Difference of Two Squares

If the sum of two numbers is multiplied by the difference of two numbers, then the product is the square of the first minus the square of the second. Thus, $(x + 5)(x - 5) = x^2 - 25$.

If an expression is the difference of two squares, then the factors are the sum and difference of the square roots of the squares. Thus, $a^2 - 64 = (a + 8)(a - 8)$.

Example 6.3
Factor:

a) $36 - a^2$
b) $x^2 - 100y^2$

Ans. a) $(6 + a)(6 - a)$
b) $(x + 10y)(x - 10y)$

Finding the Product of Two Binomials with Like Terms

Multiply: $3x + 5$ by $2x + 4$

Procedure: Solution:
1. Multiply the first
   terms together: $6x^2$
2. Add the product of the
   outer terms to the product
   of the inner terms: $12x + 10x$
3. Multiply the last terms together: 20

4. Combine the resulting terms: $6x^2 + 22x + 20$ *Ans.*

Problem 6.4
Multiply :

   a) $(2c - b)(5c - b)$ *Ans.*   a) $10c^2 - 7bc + b^2$

Factoring Trinomials

Factoring Trinomials of the Form $x^2 + bx + c$

A trinomial in the form of $x^2 + bx + c$ may or may not be factorable into binomial factors. If factoring is possible, use the following procedure:

Factor: $x^2 + 6x + 5$

Procedure: Solution:
1. Obtain the factors x Factor x^2:
   and x of x^2. Use each $x \cdot x$
   as the first term of each
   binomial:

2. Select from the factors
   of the last term c the
   pair whose sum = b:

   Factor +5:
   $(-5)(-1)$ and $(5)(1)$

   Select $(5)(1)$ since
   sum = +6 which is the
   middle term.

3. Form binomial factors
   from factors obtained in
   steps 1 and 2:

   $(x + 5)(x + 1)$ *Ans.*

Problems 6.5
Factor:

a) $x^2 + 6x - 7$
b) $x^2 + 6x + 8$
c) $x^2 + 5x + 6$
d) $x^2 - 3x - 4$

Ans.
a) $(x + 7)(x - 1)$
b) $(x + 4)(x + 2)$
c) $(x + 3)(x + 2)$
d) $(x - 4)(x + 1)$

Factoring Trinomials in the Form of $ax^2 + bx + c$

A trinomial in the form of $ax^2 + bx + c$ may or may not be factorable into binomial factors. If this is possible, use the following procedure:

Factor: $2x^2 - 11x + 5$

Procedure:
1. Factor the first term, ax^2.
   Use each as the first term
   of each binomial factor:

Solution:
Factor $2x^2$:
$(2x)(x)$

2. Select from the factors
    of the last term, c, those
    factors to be used as the
    second term of each binomial   Factor +5:
    such that the middle term, bx, results: $(-5)(-1)$ and $(+5)(+1)$

> Select $(-5)(-1)$ to obtain a
> sum of outside product $-10x$
> and inside product $-x$ equal
> to $-11x$ which is the middle term.

3. Form binomial factors from
    factors obtained in steps 1 and 2: $(2x - 1)(x - 5)$ *Ans.*

4. Test using FOIL:
    First terms multiplied together
    Outside terms multiplied together
    Inside terms multiplied together $2x^2 - 10x - 1x + 5$
    Last terms multiplied together $2x^2 - 11x + 5$

Problem 6.6
Factor :

a) $2x^2 - 7x + 5$ *Ans.*   a) $(2x - 5)(x - 1)$
b) $3x^2 + 4x - 4$ b) $(3x - 2)(x + 2)$
c) $4x^2 + 9x + 2$ c) $(4x + 1)(x + 2)$
d) $6x^2 - 7x - 5$ d) $(3x - 5)(2x + 1)$

Squaring a Binomial

The **square of a binomial is a perfect-square trinomial.** Thus, the square of $x + y$ or $(x + y)^2$ is the perfect-square trinomial $x^2 + 2xy + y^2$.

Square: $3x + 5$

☆ Note!

Square $3x + 5$ means $(3x + 5)^2$ means $(3x + 5)(3x + 5)$.

Procedure: Solution:

1. Square the first term to
   obtain the first term of the   Square $(3x)^2$
   trinomial: $= 9x^2$

2. Double the product of both
   terms to obtain the middle   Double $(3x)(+5)$
   term of the trinomial: $= 30x$

3. Square the last term to obtain   Square $+5$
   the last term of the trinomial: $= (5)^2$
                                            $= 25$

4. Combine the resulting terms: $9x^2 + 30x + 25$ *Ans.*

Factoring a Perfect-Square Trinomial

The factors of a perfect-square trinomial are two equal binomials. Thus, the factors of the perfect-square trinomial $x^2 + 2xy + y^2$ are $x + y$ and $x + y$. A perfect-square trinomial has:

1. Two terms, the first and last, which are positive perfect squares, and

2. A remaining term which is double the product of the square roots of the other two terms. This term may be positive or negative.

Thus, $x^2 + 14x + 49$ and $x^2 - 14x + 49$ are perfect-square trinomials.

Thus, $x^2 + 14x + 49 = (x + 7)^2$ and $x^2 - 14x + 49 = (x - 7)^2$.

Factor: $4x^2 - 20x + 25$

Procedure: Solution:

1. Find the principal square root of the first term and this becomes the first term of each binomial:

$$\sqrt{4x^2} = 2x$$

2. Find the principal square root of the last term and this becomes the last term of each binomial:

$$\sqrt{25} = 5$$

3. Form a binomial from results in steps 1 and 2. The answer is the binomial squared:

$$(2x - 5)^2 \quad Ans.$$

Problems 6.7
Factor.

a) $25y^2 - 10y + 1$	*Ans.* a) $(5y - 1)^2$
b) $9y^2 - 24y + 16$	b) $(3y - 4)^2$
c) $4y^2 + 28y + 49$	c) $(2y + 7)^2$
d) $25 - 10xy + x^2y^2$	d) $(5 - xy)^2$

Factoring Polynomials Completely

To factor an expression completely, continue factoring until the polynomial factors cannot be factored further. If an expression has a common monomial factor:

1. First, remove its highest common factor (HCF).

2. Then, continue factoring its polynomial until no further factors remain.

Thus, to factor $5x^2 - 5$ first factor it into $5(x^2 - 1)$, then factor it further into $5(x-1)(x+1)$.

Example 6.1
Factor completely :

    a) $3b^2 - 27$

    b) $5x^2 + 10x + 5$

    c) $6x^2 - 9x - 15$

Ans. a) 1. Remove HCF : $3(b^2 - 9)$
           2. Continue factoring : $3(b - 3)(b + 3)$ *Ans.*

    b) 1. Remove HCF : $5(x^2 + 2x + 1)$
        2. Continue factoring : $5(x + 1)^2$ *Ans.*

    c) 1. Remove HCF $3(2x^2 - 3x - 5)$
        2. Continue factoring: $3(2x - 5)(x + 1)$ *Ans.*

Chapter 7
FRACTIONS

Understanding Fractions

The **terms of a fraction** are its numerator and its denominator. Thus, in the fraction $\dfrac{8}{9}$, the terms are: 8, the numerator, and 9, the denominator.

A fraction may have a different meaning depending upon how it is used. Three of the meanings are provided below:

- *Meaning 1* — **A fraction** may mean **division.** Thus, $\dfrac{3}{4}$ may mean 3 divided by 4 or 3 ÷ 4.

- *Meaning 2* — **A fraction** may mean **ratio.** Thus, $\dfrac{3}{4}$ may mean the ratio of 3:4.

- *Meaning 3* — **A fraction** may mean **a part of a whole thing** or **a part of a group of things.** Thus, $\dfrac{3}{4}$ may mean three–fourths of a dollar or 3 out of 4 dollars.

When the numerator of a fraction is zero, the value of the fraction is zero unless the denominator is zero. Since division by zero is impossible, *any* fraction with a zero denominator has no meaning. Thus, $\dfrac{0}{4} = 0$, however $\dfrac{0}{0}$ is meaningless.

Changing Fractions to Equivalent Fractions

Equivalent fractions are fractions having the same value although they have different numerators and denominators. Thus, $\dfrac{3}{4}$, $\dfrac{30}{40}$, and $\dfrac{3x}{4x}$ are equivalent fractions.

To obtain equivalent fractions, use one of the following fraction rules:

- *Rule 1 —* The value of a fraction is not changed if its numerator and denominator are **multiplied** by the same number except zero. Thus, $\dfrac{3}{4} = \dfrac{3(10)}{4(10)} = \dfrac{30}{40}$. Both 3 and 4 are multiplied by 10.

- *Rule 2 —* The value of a fraction is not changed if its numerator and denominator are divided by the same number except zero. Thus, $\dfrac{30}{40} = \dfrac{30 \div 10}{40 \div 10} = \dfrac{3}{4}$. Both 3 and 4 are divided by 10.

Reciprocals and Their Uses

The reciprocal of a number is 1 divided by the number. Thus, the reciprocal of 5 is $\dfrac{1}{5}$. The rules regarding reciprocals are as follows:

- *Rule 1* — The fractions $\dfrac{a}{b}$ and $\dfrac{b}{a}$ are reciprocals of each other.

- *Rule 2* — The product of two reciprocals is 1. Thus, $\dfrac{a}{b}\left(\dfrac{b}{a}\right)=1$

- *Rule 3* — To divide by a number or a fraction, multiply by its reciprocal. Thus, $8\div\dfrac{2}{3}=8\times\dfrac{3}{2}.$

- *Rule 4* — To solve an equation for an unknown having a fractional coefficient, multiply both members by the reciprocal fraction. Thus, if $\dfrac{2}{3}x=10$, multiply both sides by $\dfrac{3}{2}$.

Reducing Fractions to Lowest Terms

A fraction is reduced to lowest terms when its numerator and denominator have no common factor except 1. Remember the following rules when reducing a fraction to its lowest terms:

- *Rule 1* — If two expressions are exactly alike or have the same value, their quotient is 1. Thus, $\dfrac{5ab}{5ab}=1$.

- *Rule 2* — If two binomials are negatives of each other, their quotient is −1. Thus, $\dfrac{x-y}{y-x}=\dfrac{x-y}{-1(x-y)}=-1$.

Warning!

Do not add or subtract the same number from the numerator and denominator.

Thus, $\dfrac{5}{6}$ does not equal $\dfrac{5-4}{6-4} = \dfrac{1}{2}$. Also,

$\dfrac{x}{y}$ does not equal $\dfrac{x+3}{y+3}$.

Multiplying Fractions

Multiply: $\dfrac{3}{5} \cdot \dfrac{10}{7} \cdot \dfrac{77}{6}$

Procedure:

Solution:

1. Factor the numerator and denominator:

$$\frac{3 \cdot 2 \cdot 5 \cdot 7 \cdot 11}{5 \cdot 7 \cdot 2 \cdot 3}$$

2. Cancel out factors in common to numerator and denominator:

Cancel 2, 3, 5, 7

3. Multiply remaining factors:

11   *Ans.*

Dividing Fractions

To divide by a fraction, invert the fraction and multiply.

Divide: $\dfrac{14}{x} \div 2\dfrac{1}{3}$

Procedure:	Solution:
1. Rewrite problem:	$\dfrac{14}{x} \div \dfrac{7}{3}$
2. Invert fraction:	$\dfrac{14}{x} \cdot \dfrac{3}{7}$
3. Multiply resulting fraction:	$\dfrac{6}{x}$ A*ns.*

Adding or Subtracting Fractions Having the Same Denominator

Combine: $\dfrac{2a}{15} + \dfrac{7a}{15} - \dfrac{4a}{15}$

Procedure:	Solution:
1. Keep denominator and combine numerators:	$\dfrac{2a + 7a - 4a}{15}$
2. Reduce resulting fraction:	$\dfrac{5a}{15} = \dfrac{a}{3}$ *Ans.*

Example 7.1
Combine.

a) $\dfrac{7}{x-2} - \dfrac{5+x}{x-2}$

An*s.* a) 1. Combine: $\dfrac{7-(5+x)}{x-2}$

$$\dfrac{7-5-x}{x-2}$$

2. Reduce: $\dfrac{2-x}{x-2}$

3. Simplify: $= -1$ *Ans.*

Adding or Subtracting Fractions Having Different Denominators

The first step in adding or subtracting with fractions having different denominators is to determine the Least Common Denominator or LCD. Use the rules below to determine the LCD:

• *Rule 1* — If no two denominators have a common factor, find the LCD by multiplying all the denominators. Thus, $3ax$ is the LCD of $\dfrac{1}{3}, \dfrac{1}{a}$, and $\dfrac{1}{x}$.

• *Rule 2* — If two of the denominators have a common factor, find the LCD by multiplying the common factor by the remaining factors. Thus, for $\dfrac{1}{3xy}$ and $\dfrac{1}{5xy}$, the LCD, $15xy$, is obtained by multiplying the common factor xy by the remaining factors, 3 and 5.

- *Rule 3* — If there is a common variable factor with more than one exponent, use its highest exponent in the LCD. Thus, $3y^5$

  is the LCD of $\dfrac{1}{3y}, \dfrac{1}{y^2}$, and $\dfrac{1}{y^5}$.

Combine: $\dfrac{s^2}{9r^2} - \dfrac{s^3}{12r^3}$

Procedure:	Solution:

1. Find the LCD:

$$LCD = 36r^3$$

2. Change each fraction to an equivalent fraction whose denominator is the LCD by multiplying by a fraction that is equal to 1:

$$\frac{s^2}{9r^2} \cdot \frac{4r}{4r} = \frac{4rs^2}{36r^3}$$

$$\frac{s^3}{12r^3} \cdot \frac{3}{3} = \frac{3s^3}{36r^3}$$

3. Combine fractions having the same denominator and reduce, if necessary:

$$\frac{4rs^2}{36r^3} - \frac{3s^3}{36r^3}$$

$$= \frac{4rs^2 - 3s^3}{36r^3} \quad Ans.$$

Example 7.2
Combine:

a) $\dfrac{3x}{x-2} + \dfrac{5x}{x+2}$

Ans. a) 1. Find LCD by multiplying the denominators:

LCD is $(x - 2)(x + 2)$

2. Muliply by 1.

$$\frac{3x}{x-2} \cdot \left(\frac{x+2}{x+2}\right) + \frac{5x}{x+2} \cdot \left(\frac{x-2}{x-2}\right)$$

3. Change denominator:

$$\frac{3x^2 + 6x}{(x+2)(x-2)} + \frac{5x^2 - 10x}{(x+2)(x-2)}$$

4. Combine and reduce:

$$\frac{8x^2 - 4x}{(x+2)(x-2)} \ Ans.$$

Simplifying Complex Fractions

A complex fraction is a fraction containing at least one other fraction within it. Thus,

$$\frac{\frac{3}{4}}{\frac{2}{2}}, \frac{5}{2}, \quad \frac{x+\frac{1}{2}}{x-\frac{1}{4}} \quad \text{and} \quad \text{are complex fractions.}$$

To simplify a complex fraction use one of the two methods presented below:

The LCD–Multiplication Method

Simplify:
$$\frac{x - \frac{1}{3}}{\frac{3}{5} + \frac{7}{10}}$$

Procedure:	Solution:
1. Find the LCD of all of the fractions in the complex fraction:	$LCD = 30$
2. Multiply both numerator and denominator by LCD, and reduce, if necessary:	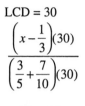
	$= \dfrac{30x-10}{39}$ *Ans.*

The Combining–Division Method

Simplify: $\dfrac{1+\dfrac{2}{y}}{1-\dfrac{4}{y^2}}$

Procedure:	Solution:
1. Combine terms of numerator:	$1+\dfrac{2}{y}=\dfrac{y}{y}+\dfrac{2}{y}=\dfrac{y+2}{y}$
2. Combine terms of denominator:	$1-\dfrac{4}{y^2}=\dfrac{y^2}{y^2}-\dfrac{4}{y^2}=\dfrac{y^2-4}{y^2}$
3. Divide new numerator	$\dfrac{y+2}{y}\div\dfrac{y^2-4}{y^2}$
by new denominator:	$=\dfrac{y+2}{y}\cdot\dfrac{y^2}{y^2-4}$
4. Simplify:	$=\dfrac{y}{y-2}$ *Ans.*

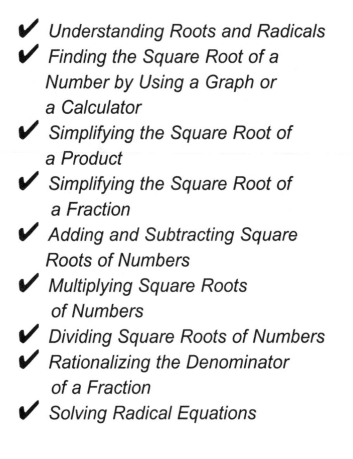

Chapter 8
ROOTS AND RADICALS

IN THIS CHAPTER:

✔ *Understanding Roots and Radicals*
✔ *Finding the Square Root of a Number by Using a Graph or a Calculator*
✔ *Simplifying the Square Root of a Product*
✔ *Simplifying the Square Root of a Fraction*
✔ *Adding and Subtracting Square Roots of Numbers*
✔ *Multiplying Square Roots of Numbers*
✔ *Dividing Square Roots of Numbers*
✔ *Rationalizing the Denominator of a Fraction*
✔ *Solving Radical Equations*

Understanding Roots and Radicals

The square root of a number is one of its two equal factors. Thus, +5 is the **principal square root** of 25. Also, –5 is another square root of 25 since (–5)(–5) = 25. The symbol $\sqrt{}$ indicates the principal square root of a number. To indicate the negative square root of a number, place the minus sign before the root symbol. Thus, $-\sqrt{25} = -5$

Radical, Radical Sign, Radicand, Index

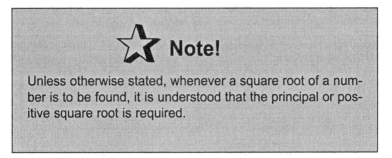

★ Note!

Unless otherwise stated, whenever a square root of a number is to be found, it is understood that the principal or positive square root is required.

- **A radical** is an indicated root of a number or an expression. Thus, $\sqrt{5}, \sqrt[3]{8x}, and \sqrt[4]{7x^3}$ are radicals.

- The symbols, $\sqrt{}, \sqrt[3]{}$, and $\sqrt[4]{}$ are **radical signs.**

- The **radicand** is the number or expression under the radical sign. Thus, 80 is the radicand of $\sqrt{80}$.

- The **index of a root** is the small number written above and to the left of the radical sign $\sqrt{}$. Thus, $\sqrt[3]{8}$ indicates the third root of 8 or the cube root of 8. In square roots, the index 2 is not indicated but is understood.

- **The cube root of a number** is one of its three equal factors. Thus, –3 is a cube root of –27 since (–3)(–3)(–3) = –27.

Finding the Square Root of a Number by Using a Graph or a Calculator

Approximate square roots of numbers can be obtained by using a graph of $x = \sqrt{y}$. To find the square root of a number graphically follow the procedure below using the graph of the function :

Example 8.1

Find graphically: $\sqrt{27}$

Procedure: Solution:

1. Find the number
   on the *y–axis*:   Find 27 on the *y–axis*.

2. From the number
   proceed horizontally   From 27 follow the horizontal
   to the curve: line to point *A* on the curve.

3. From the curve
   proceed to the *x–axis*:   From *A* follow the vertical line.

4. Read the approximate
   square root value on the
   x–axis: Read 5.2 on the *x*–axis.

$$\sqrt{27} = 5.2 \quad Ans.$$

Approximate square roots of numbers can be obtained by using a calculator. To find the principal square root of a number, solutions using the calculator will be more precise than those on a graph. Answers which are nonrepeating decimals are called **irrational numbers**. A **rational number** is one that can be expressed as the quotient or ratio of two integers. Therefore, an irrational number is one that cannot be expressed as the ratio of two integers.

Simplifying the Square Root of a Product

$$\sqrt{ab} = \sqrt{a} \cdot \sqrt{b} \quad \text{and} \quad \sqrt{abc} = \sqrt{a} \cdot \sqrt{b} \cdot \sqrt{c} .$$

- *Rule 1* — The square root of a product of two or more numbers equals the product of the separate square roots of these roots. Thus, $\sqrt{3600} = \sqrt{(36)(100)} = \sqrt{36}\sqrt{100} = 60$.

- *Rule 2* — To find the square root of a number raised to a power, keep the base and take one-half of the exponent. Thus, $\sqrt{x^6} = x^3$ since one-half of 6 is 3.

- *Rule 3* — To find the square root of the product of powers, keep each base and take one-half of the exponents. Thus, $\sqrt{x^2 y^4} = xy^2$ since $\sqrt{x^2}\sqrt{y^4} = xy^2$. One-half of 2 is 1 and one-half of 4 is 2.

Example 8.2
Simplify.

   a) $\sqrt{112}$

   b) $3\sqrt{450}$

Ans. a) $\sqrt{16 \cdot 7}$

   $4\sqrt{7}$ *Ans.*

   b) $3\sqrt{9 \cdot 25 \cdot 2}$

   $3\sqrt{9} \cdot \sqrt{25} \cdot \sqrt{2}$
   $3 \cdot 3 \cdot 5 \cdot \sqrt{2}$
   $45\sqrt{2}$.

Example 8.3

Using $\sqrt{2} = 1.414$, evaluate to the nearest tenth.

   a) $\sqrt{72}$

Ans. a) $\sqrt{72} = \sqrt{36 \cdot 2}$
   $= 6\sqrt{2}$

   $= 6(1.414)$
   $= 8.5$

Simplifying the Square Root of a Fraction

$$\sqrt{\frac{a}{b}} = \frac{\sqrt{a}}{\sqrt{b}}$$

• *Rule* — The square root of a fraction equals the square root of the numerator divided by the square root of the denominator.

Thus, $\sqrt{\dfrac{25}{64}} = \dfrac{\sqrt{25}}{\sqrt{64}} = \dfrac{5}{8}$.

To simplify the square root of a fraction, whose denominator is not a perfect square, **change the fraction to an equivalent fraction,** which has a denominator that is the smallest perfect square. Thus,

$$\sqrt{\frac{1}{8}} = \sqrt{\frac{2}{16}} = \frac{\sqrt{2}}{\sqrt{16}} = \frac{\sqrt{2}}{4} \text{ or } \frac{1}{4}\sqrt{2}.$$

Problem 8.1
Simplify:

a) $6\sqrt{\dfrac{7}{9}}$

b) $\sqrt{\dfrac{a}{3}}$

Ans. a) $2\sqrt{7}$

b) $\dfrac{\sqrt{3a}}{3}$

Adding and Subtracting Square Roots of Numbers

Like radicals are radicals having the same index and the same radicand. Thus:

Like Radicals	Unlike Radicals
$5\sqrt{3}$ and $2\sqrt{3}$ $5\sqrt{x}$ and $2\sqrt{x}$	$5\sqrt{3}$ and $2\sqrt{5}$ $5\sqrt{x}$ and $2\sqrt{y}$

- *Rule* — **To combine (add or subtract) like radicals,** keep the common radical and combine their coefficients. Thus,

$$5\sqrt{3} + 2\sqrt{3} - 4\sqrt{3} = (5 + 2 - 4)\sqrt{3} = 3\sqrt{3}$$

Remember

Combining like radicals involves the same process as combining like terms. Hence, to combine $5\sqrt{3}$ and $2\sqrt{3}$, think of combining 5x and 2x when $x = \sqrt{3}$.

Multiplying Square Roots of Numbers

$$\sqrt{a}\sqrt{b} = \sqrt{ab} \text{ and } \sqrt{a}\sqrt{b}\sqrt{c} = \sqrt{abc}$$

- *Rule 1* — The product of the square roots of two or more nonnegative numbers equals the square root of their product.

  Thus, $\sqrt{2}\sqrt{3}\sqrt{6} = \sqrt{36} = 6$.

- *Rule 2* — The square of the square root of a number equals the number. Thus, $\sqrt{7}\sqrt{7} = \sqrt{49} = 7$. In general,

  $(\sqrt{x})^2 = x$.

Example 8.4

Multiply: $\dfrac{2}{3}\sqrt{2x} \cdot 6\sqrt{2y}$

Procedure:	Solution:
1. Multiply coefficients and radicals, separately:	$\dfrac{2}{3} \cdot 6\sqrt{2x}\sqrt{2y}$
2. Multiply the resulting products:	$4\sqrt{4xy}$
3. Simplify, if possible:	$4 \cdot 2\sqrt{xy}$
	$= 8\sqrt{xy}$ *Ans.*

Dividing Square Roots of Numbers

$$\frac{\sqrt{a}}{\sqrt{b}} = \sqrt{\frac{a}{b}}$$

- *Rule* — The square root of a number divided by the square root of another number equals the square root of their quotient. Thus, $\dfrac{\sqrt{6}}{\sqrt{2}} = \sqrt{\dfrac{6}{2}} = \sqrt{3}$.

Example 8.5

Divide: $\dfrac{14\sqrt{40}}{2\sqrt{5}}$

Procedure:	Solution:
1. Divide coefficients and radicals separately:	$\dfrac{14\sqrt{40}}{2\sqrt{5}}$
2. Multiply and simplify:	$7\sqrt{8}$
3. Simplify, if possible:	$\begin{aligned} &7\sqrt{4\cdot 2} \\ =\,&14\sqrt{2} \quad Ans. \end{aligned}$

Problem 8.2
Divide:

a) $\dfrac{6\sqrt{x^4}}{3\sqrt{x}}$ *Ans.* a) $2x\sqrt{x}$

b) $\dfrac{\sqrt{50}+\sqrt{98}}{\sqrt{2}}$ b) 12

Rationalizing the Denominator of a Fraction

To rationalize the denominator of a fraction is to change the denominator to a rational number. A rational number is one that can be expressed as the quotient or ratio of two integers. (Note the word ratio in the word rational.) To do this when the denominator is a monomial, multiply both terms of the fraction by a radical that will make the radicand of the denominator the smallest perfect square. Thus, to rationalize the denominator of $\dfrac{4}{\sqrt{8}}$, multiply by $\dfrac{\sqrt{2}}{\sqrt{2}}$. Rationalizing the denominator simplifies the evaluation of the equation. Thus,

$$\dfrac{4}{\sqrt{8}} \cdot \dfrac{\sqrt{2}}{\sqrt{2}} = \dfrac{4\sqrt{2}}{\sqrt{16}} = \sqrt{2} \approx 1.414$$

Problem 8.3
Rationalize:

a) $\dfrac{7}{\sqrt{7}}$ *Ans.* a) $\sqrt{7}$

b) $\dfrac{1}{\sqrt{c^3}}$ b) $\dfrac{\sqrt{c}}{c^2}$

Solving Radical Equations

Radical equations are equations in which the variable is included in a radicand. Thus, $2\sqrt{x} + 5 = 9$ is a radical equation, but $2x + \sqrt{5} = 9$ is not a radical equation.

Example 8.6
Solve: $\sqrt{2x} + 5 = 9$

Procedure:
1. Isolate the term containing the radical:

Solution:
$$\sqrt{2x} + 5 = 9$$
$$\sqrt{2x} = 4$$

2. Square both sides: $\qquad\qquad\qquad\qquad$ $2x = 16$

3. Solve for the unknown: $\qquad\qquad\qquad$ $x = 8$ $\quad$ *Ans.*

4. Check the answer in the
   original equation: $\qquad\qquad\qquad$ $\sqrt{2 \cdot 8} + 5 = 9$

   $$9 = 9 \quad \text{(Yes)}$$

An **extraneous root** of an equation is a value of the unknown that solves an equation within the problem, but does not solve the original equation. Thus, if $1 - \sqrt{x} = 2$ is solved using the procedure above, x will equal 1, but 1 does not solve the original equation.

Problem 8.4
Solve and check:

$\quad$ a) $\sqrt{7x+5} = 3$ $\qquad\qquad$ *Ans.* $\quad$ a) $\dfrac{4}{7}$

$\quad$ b) $\sqrt{3x+4} - 2 = 3$ $\qquad\qquad\qquad$ b) 7

Problem 8.5
Solve for y:

$\quad$ a) $\sqrt{2y-5} = x$ $\qquad\qquad$ *Ans.* $\quad$ a) $y = \dfrac{x^2 + 5}{2}$

Solve for x:

$\quad$ b) $\sqrt{\dfrac{x}{2}} = y$ $\qquad\qquad\qquad$ b) $x = 2y^2$

Chapter 9
QUADRATIC EQUATIONS IN ONE VARIABLE

IN THIS CHAPTER:

- ✔ *Understanding Quadratic Equations in One Unknown*
- ✔ *Solving Quadratic Equations by Factoring*
- ✔ *Solving Incomplete Quadratic Equations*
- ✔ *Solving a Quadratic Equation by Completing the Square*
- ✔ *Solving a Quadratic Equation by the Quadratic Formula*
- ✔ *Solving Quadratic Equations Graphically*

Understanding Quadratic Equations in One Unknown

A **quadratic equation** in one unknown is an equation in which the highest power of the unknown is the second. Thus, $2x^2 + 3x - 5 = 0$ is a quadratic equation in x.

Standard Quadratic Equation Form

The form of a standard quadratic equation in one unknown is:

$$ax^2 + bx + c = 0$$

where a, b, and c represent known numbers and x represents the unknown number. The number a cannot equal zero. Thus, $3x^2 - 5x + 6 = 0$ is in standard form.

Follow the steps outlined below to transfer a quadratic equation into standard form:

- *Step 1* — **Remove parentheses.**

  Thus, $x(x + 1) - 5 = 0$ becomes $x^2 + x - 5 = 0$.

- *Step 2* — **Clear of fractions.**

  Thus, $x - 4 + \dfrac{3}{x} = 0$ becomes $x^2 - 4x + 3 = 0$.

- *Step 3* — **Remove radical signs.**

  Thus, $\sqrt{x^2 - 3x} = 2$ becomes $x^2 - 3x - 4 = 0$.

- *Step 4* — **Collect like terms.**

  Thus, $x^2 + 7x = 2x + 6$ becomes $x^2 + 5x - 6 = 0$.

Solving Quadratic Equations by Factoring

- *Rule 1* — Every quadratic equation has two roots. Thus, $x^2 = 9$ has two roots, 3 and −3; that is, $x = \pm 3$.

- *Rule 2* — If the product of two factors is zero, then one or both of the factors must equal zero. Thus,
  - In $5(x-3) = 0$, the factor $x - 3 = 0$.
  - In $(x - 2)(x - 3) = 0$, either $(x - 2)$ or $(x - 3) = 0$ or both.
  - In $(x - 3)(x - 3) = 0$, both factors $x - 3 = 0$.

Example 9.1
Solve: $x\,(x - 4) = 5$

Procedure:	Solution:
1. Express in form $ax^2 + bx + c = 0$:	$x^2 - 4x = 5$ $x^2 - 4x - 5 = 0$
2. Factor $ax^2 + bx + c = 0$:	$(x - 5)(x + 1) = 0$
3. Let each factor equal zero:	$(x - 5) = 0$ and $(x + 1) = 0$
4. Solve each resulting equation:	$x = 5$ and $x = -1$ *Ans.*
5. Check each root in the problem:	$x\,(x - 4) = 5$:
	$5(5 - 4) = 5$ (Yes) $-1(-1 - 4) = 5$ (Yes)

Problem 9.1
Solve by factoring:

a) $x^2 + 9x + 20 = 0$
b) $x^2 - x = 6$

Ans. a) $x = -4$ or -5
    b) $x = 3$ or -2

c) $x^2 - 49 = 0$ c) $x = 7$ or -7
d) $x^2 - 11 = 25 - 5x$ d) $x = 4$ or -9

Solving Incomplete Quadratic Equations

An **incomplete quadratic equation** in one unknown lacks one of the following:

1. The term containing the first power of the unknown as in $x^2 - 4 = 0$.

2. The constant term as in $x^2 - 4x = 0$.

Example 9.2
Solve: $2(x^2 - 8) = 11 - x^2$

Procedure:	Solution:

1. Express in form $ax^2 = k$,
   where k is a constant: $2x^2 - 16 = 11 - x^2$
                                    $3x^2 = 27$

2. Divide both sides by a: $x^2 = 9$

3. Take the square root of both sides: $x = 3$ and -3 *Ans.*

4. Check each root in the problem: $2(x^2 - 8) = 11 - x^2$:
                                       $2(3^2 - 8) = 11 - 3^2$
                                       $2((-3)^2 - 8) = 11 - (-3)^2$

Problem 9.2
Solve for x in the incomplete quadratics:

a) $x^2 - 64 = 0$ *Ans.* a) $x = +8$ or -8
b) $3x^2 = 27x$ b) $x = 0$ or $+9$

   c) $-3(x^2 - 8) = 24 + 15x$ c) $x = 0$ or -5

   d) $8x^2 - 800 = 0$ d) $x = +10$ or -10

Solving a Quadratic Equation by Completing the Square

The square of a binomial is a perfect-square trinomial. Thus, $x^2 + 6x + 9$ is the perfect-square trinomial of $x + 3$.

- *Rule* —   If x^2 is the first term of a perfect-square trinomial and the term in x is also given, the last term may be found by squaring one–half the coefficient of x.

Thus, in $x^2 + 6x$, the number 9 is needed to complete the perfect-square trinomial $x^2 + 6x + 9$. The last term, 9, is found by squaring half of 6, or 3.

Example 9.3

Solve by completing the square: $x^2 + 6x - 7 = 0$

Procedure:

Solution:

1.   Express the equation in the form of $x^2 + px = q$:

                                 Change $x^2 + 6x - 7 = 0$

                                 to $x^2 + 6x = 7$

2.   Square one-half the coefficient of x and add to both sides:

                                   The square of

$$\frac{1}{2}(6) = 3^2 = 9$$

                                   Add 9 to get:

                                   $x^2 + 6x + 9 = 16$

3.   Replace the perfect trinomial square by its binomial squared:

                                   $(x + 3)^2 = 16$

4.   Take the square root

of both sides: $x + 3 = 4$ or
$x + 3 = -4$

5. Solve the two resulting
equations: $x = 1$ or $x = -7$ *Ans.*

Problem 9.3
Complete each perfect-square trinomial and state its binomial squared:

a) $x^2 + 14x + ?$ *Ans.*   a) Add 49, $(x + 7)^2$
b) $x^2 - 20x + ?$ b) Add 100, $(x - 10)^2$

c) $x^2 + x + ?$ c) Add 1/4, $(x + \dfrac{1}{2})^2$

d) $x^2 - 2.4x + ?$ d) Add 1.2^2, $(x + 1.2)^2$

Problem 9.4
Solve by completing the square :

a) $x^2 - 4x = 5$ *Ans.*   a) $5, -1$
b) $x^2 + 14x - 32 = 0$ b) $2, -16$
c) $x(x - 5) = -4$ c) $4, 1$

d) $5x^2 - 4x = 33$ d) $3, -\dfrac{11}{5}$

Solving a Quadratic Equation by Quadratic Formula

Quadratic Formula: If $ax^2 + bx + c = 0$, then by completing the square,

it may be proven that $x = \dfrac{-b \pm \sqrt{b^2 - 4ac}}{2a}$

To Solve a Quadratic Equation by the Quadratic Formula

Example 9.4
Solve $2x^2 + 4x = 3$ by the quadratic formula.

<u>Procedure:</u>

1. Express in form of
   $ax^2 + bx + c = 0$.

2. State the values of a, b, and c:

3. Substitute the values:

4. Solve for x

5. Simplify

<u>Solution:</u>

1. $2x^2 + 4x = 3$
   $2x^2 + 4x - 3 = 0$

2. $a = 2, b = 4, c = -3$

3. $x = \dfrac{-4 \pm \sqrt{4^2 - 4(2)(-3)}}{2(2)}$.

4. $x = \dfrac{-4 \pm \sqrt{40}}{4}$

5. $x = \dfrac{-2 \pm \sqrt{10}}{2}$ *Ans.*

Example 9.5
Find the roots of $3x^2 + 4x - 5 = 0$ in simplified form.

$$x = \frac{-b \pm \sqrt{b^2 - 4ac}}{2a}$$

$$x = \frac{-4 \pm \sqrt{4^2 - 4(3)(-5)}}{2(3)}$$

$$x = \frac{-4 \pm \sqrt{16 + 60}}{6}$$

$$x = \frac{-4 \pm \sqrt{76}}{6}.$$

$$x = \frac{-2 \pm \sqrt{19}}{3} \text{ *Ans.*}$$

Problem 9.5
Solve for x in simplest radical form.

a) $x^2 - 2x - 10 = 0$ *Ans.*   a) $x = 1 \pm \sqrt{11}$

b) $3x^2 - 5x = 10$ b) $x = \dfrac{5 \pm \sqrt{145}}{6}$

Problem 9.6
Solve for x, to the nearest tenth, using the simplified radical form. Find the square root to two decimal places by using the calculator.

a) $3x^2 + 4x - 5 = 0$ *Ans.* $x = \dfrac{-4 \pm \sqrt{76}}{6}$

$$x = \dfrac{-4 \pm 8.72}{6}$$

$$x = 0.8 \text{ or } x = -2.1$$

Solving Quadratic Equations Graphically

Example 9.6
Solve graphically: $x^2 - 5x + 4 = 0$. See Figure 9-1.

Procedure:	Solution:
1) Express in form $ax^2 + bx + c = 0$	1. $x^2 - 5x + 4 = 0$
2) Graph the curve $y = ax^2 + bx + c$	2. See Figure 9-1

GRAPH of $y = x^2 - 5x + 4 = 0$ (parabola)

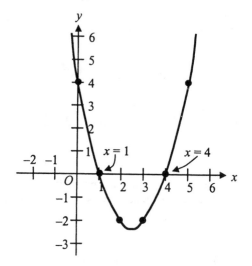

Figure 9-1

3. Find where $y = 0$ intersects $y = ax^2 + bx + c = 0$. The values of x at the points of intersection are the roots of $x^2 - 5x + 4 = 0$.

3. $x = 1$ and $x = 4$ *Ans.*

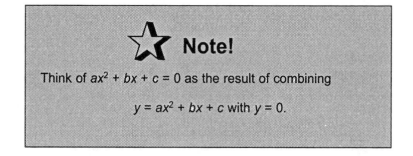

⭐ Note!

Think of $ax^2 + bx + c = 0$ as the result of combining

$$y = ax^2 + bx + c \text{ with } y = 0.$$

The following sketches are graphs showing the relationship of a parabola and the x–axis.

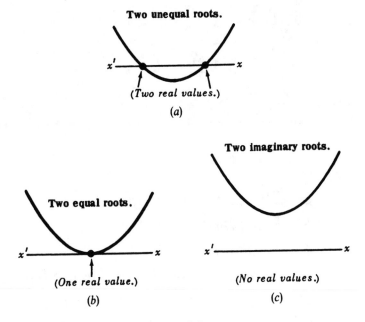

Figure 9-2

Chapter 10
INTRODUCTION TO GEOMETRY

IN THIS CHAPTER:

✔ *The Law of Pythagoras*
✔ *Proportions: Equal Ratios*
✔ *Similar Triangles*
✔ *Understanding Trigonometric Ratios*
✔ *Solving Trigonometry Problems*
✔ *Understanding Congruent Triangles*
✔ *Geometry Formulas*

The Law of Pythagoras

The Law of Pythagoras states that in a right triangle, the square of the hypotenuse equals the sum of the squares of the two legs. Thus, in triangle ABC, if C is a right angle, then $c^2 = a^2 + b^2$ and by transposition, $a^2 = c^2 - b^2$ and $b^2 = c^2 - a^2$.

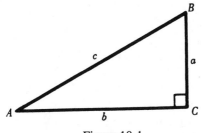

Figure 10-1

Important!

In a triangle, the small letter for a side should agree with the capital letter for the vertex of the angle opposite that side.

Thus, side *a* is opposite angle A, etc.

Example 10.1
Find the missing side in the right triangle shown:

Figure 10-2

a) $a = 12, b = 9$
b) $c = 10, a = 8$

Ans. a) $c^2 = a^2 + b^2$
$c^2 = 12^2 + 9^2$
$c^2 = 144 + 81 = 225$
$c = 15$ *Ans.*

b) $b^2 = c^2 - a^2$
$b^2 = 10^2 - 8^2$
$b^2 = 100 - 64 = 36$
$b = 6$ *Ans.*

The Law of Pythagoras can also be used to find the distance between two points on a graph. If d is the distance between $P_1(x_1, y_1)$ and $P_2(x_2, y_2)$ then:

$$d^2 = (x_2 - x_1)^2 + (y_2 - y_1)^2$$

Example 10.2
  a) Using the Law of Pythagoras, find the distance from (2, 5) to (6, 8):

Ans. a) 1. $d^2 = (x_2 - x_1)^2 + (y_2 - y_1)^2$
  2. $d^2 = (6 - 2)^2 + (8 - 5)^2$
  3. $d^2 = (4)^2 + (3)^2$
  4. $d^2 = 16 + 9 = 25$
  5. $d = 5$ *Ans.*

Proportions: Equal Ratios

A **proportion** is an equality of two ratios. Thus, $2:5 = 4:10$ or $\dfrac{2}{5} = \dfrac{4}{10}$ is a proportion.

Rule!

If *a:b* = *c:d*, then *ad* = *bc*.

Example 10.3

Solve for *x:* $\dfrac{x}{2x-3} = \dfrac{3}{5}$

Procedure:	Solution:
1. Cross-multiply	$5x = 3(2x - 3)$
2. Solve the equation.	$5x = 6x - 9$
	$x = 9$ *Ans.*

Example 10.4

a) The numerator of a fraction is 5 less than the denominator. If the numerator is doubled and the denominator is increased by 7, the value of the resulting fraction is $\dfrac{2}{3}$. Find the original fraction.

Ans. a)

1. Let x = denominator of the original fraction and $x - 5$ the numerator of the original fraction.

2. Then $\dfrac{2(x-5)}{x+7} = \dfrac{2}{3}$.

3. Cross-multiply:

$$6x - 30 = 2x + 14$$
$$4x = 44$$
$$x = 11$$

4. Thus the original fraction = $\dfrac{6}{11}$. *Ans.*

Similar Triangles

Similar polygons have the same shape. Thus, if ΔI and ΔI' are similar, then they have the same shape although they need not have the same size.

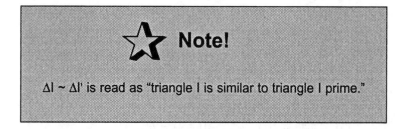

☆ **Note!**

ΔI ~ ΔI' is read as "triangle I is similar to triangle I prime."

In the diagram in Figure 10-3, note how the sides and angles having the same relative position are designated by using the same letters and primes. **Corresponding sides or angles** are those having the same relative position.

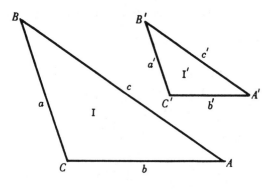

Figure 10-3

• *Rule* —   If two triangles are similar:

1. Their corresponding angles are congruent.

Thus, if $\Delta I \sim \Delta I'$ in Figure 10-4,

Then $\angle A \cong \angle A'$,

$\angle B \cong \angle B'$

$\angle C \cong \angle C'$

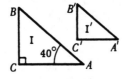

Figure 10-4

2.   The ratios of their corresponding sides are equal.

Thus, if $\Delta I \sim \Delta I'$ in Figure 10-5,

Then $c = 15$ since $\dfrac{c}{5} = \dfrac{9}{3}$

and $b = 12$ since $\dfrac{b}{4} = \dfrac{9}{3}$.

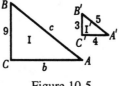

Figure 10-5

Understanding Trigonometric Ratios

Trigonometry means "measurement of triangles." Thus, trigonometry is the study of how to find the measurement of angles and lengths of sides of triangles using ratios. In trigonometry, the following ratios are used in a right triangle such as that in Figure 10-6 to relate the sides and either acute angle:

1. Tangent ratio, abbreviated tan
2. Sine ratio, abbreviated sin.
3. Cosine ratio, abbreviated cos.

Trigonometry Rules and Formulas

The table in Figure 10-7 summarizes the basic rules of trigonometry.

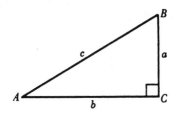

Figure 10-6

Rules	Formulas
1. The tangent of an acute angle equals the leg opposite the angle divided by the leg adjacent to the angle.	$\tan A = \dfrac{\text{leg opp. } A}{\text{leg adj. } A} = \dfrac{a}{b}$ $\tan B = \dfrac{\text{leg opp. } B}{\text{leg adj. } B} = \dfrac{b}{a}$
2. The sine of an acute angle equals the leg opposite the angle divided by the hypotenuse.	$\sin A = \dfrac{\text{leg opp. } A}{\text{hyp.}} = \dfrac{a}{c}$ $\sin B = \dfrac{\text{leg opp. } B}{\text{hyp.}} = \dfrac{b}{c}$
3. The cosine of an acute angle equals the leg adjacent to the angle divided by the hypotenuse.	$\cos A = \dfrac{\text{leg adj. } A}{\text{hyp.}} = \dfrac{b}{c}$ $\cos B = \dfrac{\text{leg adj. } B}{\text{hyp.}} = \dfrac{a}{c}$

Figure 10-7

Solving Trigonometry Problems

Finding Sides and Angles Using Trigonometric Ratios:

Example 10.5

    a) Find $m\angle A$, to the nearest degree, of the triangle in Figure 10-8.

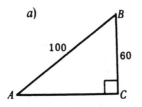

Figure 10-8

Ans. a) $\sin A = \dfrac{60}{100}$

Since $\sin 37° = .6018$ is the nearest sine value,

$m\angle A \approx 37°$.

Example 10.6
a) Find x in the triangle in Figure 10-9.

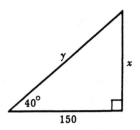

Figure 10-9

Ans. a) Since $\tan 40° = \dfrac{x}{150}$,

$x = 150 \tan 40°$ *Ans.*

Example 10.7
a) An aviator takes off at A and ascends at a fixed angle of 22° with level or horizontal ground. After flying 3000 yd., find the altitude of the plane to the nearest 10 yd. See Figure 10-10.

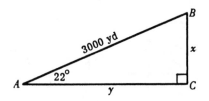

Figure 10-10

Ans. a) Let x = altitude of plane in yards.

Since sin $22° = \dfrac{x}{3000}$

$x = 3000$ sin $22°$

$x = 3000(0.3746) = 1123.8$
or 1120 to the nearest 10 yd.

Understanding Congruent Triangles

Congruent triangles have exactly the same size and shape. If triangles are congruent:

1. Their corresponding sides are congruent.

2. Their corresponding angles are congruent.

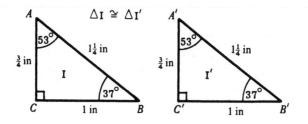

Figure 10-11

- *Rule 1* — [SSS = SSS]
   Two triangles are **congruent** if three sides of one triangle are congruent to three sides of the other.

   Thus, ΔII ≅ ΔII' (Figure 10-12) since:

1. DE = D'E' = 4 inches
2. DF = D'F' = 5 inches
3. EF = E'F' = 6 inches

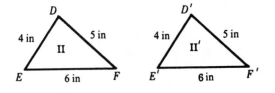

Figure 10-12

• *Rule 2 — [ASA = ASA]*

Two triangles are **congruent** if two angles and the included side of one triangle are congruent, respectively, to two angles and the included side of the other.

Thus, ΔIII ≅ ΔIII' (Figure 10-13) since:

1. ∠H ≅ ∠H' = 85°
2. ∠J ≅ ∠J' = 25°
3. HJ = H'J' = 10 inches

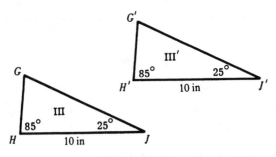

Figure 10-13

• *Rule 3* — [SAS = SAS]

Two triangles are **congruent** if two sides and the included angle of one triangle are congruent, respectively, to two sides and the included angle of the other.

Thus, ΔIV ≅ ΔIV′ (Figure 10-14) since:

1. KL = K′L′ = 9 inches
2. ∠L ≅ ∠L′ = 55°
3. LM = L′M′ = 12 inches

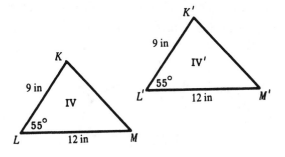

Figure 10-14

Geometry Formulas

Formulas for Perimeters and Circumference

The **perimeter of a polygon** is the distance around it. Thus, the perimeter *p* of the triangle shown is the sum of the lengths of its three sides; that is,

$$p = a + b + c.$$

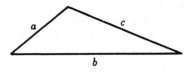

Figure 10-15

The **circumference of a circle** is the distance around it. For any circle, the circumference c is π times the diameter d; that is, $c = \pi d$. See Figure 10-16.

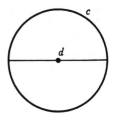

Figure 10-16

Area Formulas Using A for Area of Figure

1. **Rectangle:** $A = bh$ 2. **Parallelogram:** $A = bh$

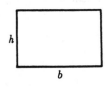   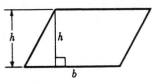

Figure 10-17   Figure 10-18

3. Triangle: $A = \dfrac{1}{2}bh$

4. Square: $A = s^2$

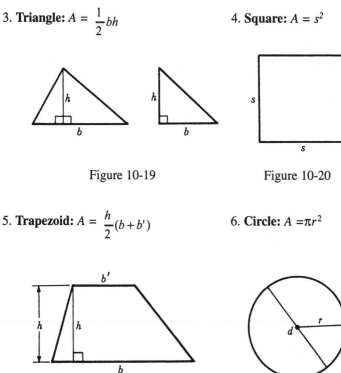

Figure 10-19

Figure 10-20

5. Trapezoid: $A = \dfrac{h}{2}(b + b')$

6. Circle: $A = \pi r^2$

Figure 10-21

Figure 10-22

Formulas for Volumes

Here we use V for the volume of the solid, B for the area of a base, and h for the distance between the bases or between the vertex and a base:

1. Rectangular solid: $V = lwh$ **2. Pyramid:** $V = \dfrac{1}{3} Bh$

3. **Prism:** $V = Bh$ 4. **Cone:** $V = \dfrac{1}{3} Bh$

5. **Cylinder:** $V = Bh$ 6. **Sphere:** $V = \dfrac{4}{3}\pi \, r^3$

7. **Cube:** $V = e^3$

Example 10.8

Find the volume, to the nearest integer, of

    a) a sphere with a radius of 10

Solution: (Let $\pi = 3.14$)

    a) $V = \dfrac{4}{3}\pi \, r^3$

        $V = \dfrac{4}{3}(3.14)(10)^3$

    Ans: 4187 in^3

Index